Medicine and Biomedical Sciences in Modern History

Series Editors

Carsten Timmermann
University of Manchester
United Kingdom

Michael Worboys
University of Manchester
United Kingdom

The aim of this series is to illuminate the development and impact of medicine and the biomedical sciences in the modern era. The series was founded by the late Professor John Pickstone, and its ambitions reflect his commitment to the integrated study of medicine, science and technology in their contexts. He repeatedly commented that it was a pity that the foundation discipline of the field, for which he popularized the acronym 'HSTM' (History of Science, Technology and Medicine) had been the history of science rather than the history of medicine. His point was that historians of science had too often focused just on scientific ideas and institutions, while historians of medicine always had to consider the understanding, management and meanings of diseases in their socio-economic, cultural, technological and political contexts. In the event, most of the books in the series dealt with medicine and the biomedical sciences, and the changed series title reflects this. However, as the new editors we share Professor Pickstone's enthusiasm for the integrated study of medicine, science and technology, encouraging studies on biomedical science, translational medicine, clinical practice, disease histories, medical technologies, medical specialisms and health policies.

The books in this series will present medicine and biomedical science as crucial features of modern culture, analysing their economic, social and political aspects, while not neglecting their expert content and context. Our authors investigate the uses and consequences of technical knowledge, and how it shaped, and was shaped by, particular economic, social and political structures. In re-launching the Series, we hope to build on its strengths but extend its geographical range beyond Western Europe and North America.

Medicine and Biomedical Sciences in Modern History is intended to supply analysis and stimulate debate. All books are based on searching historical study of topics which are important, not least because they cut across conventional academic boundaries. They should appeal not just to historians, nor just to medical practitioners, scientists and engineers, but to all who are interested in the place of medicine and biomedical sciences in in modern history

More information about this series at
http://www.springer.com/series/15183

Helen Valier

A History of Prostate Cancer

Cancer, Men and Medicine

palgrave
macmillan

Helen Valier
University of Houston
Texas, USA

ISBN 978-1-349-73938-7   ISBN 978-1-137-56595-2   (eBook)
DOI 10.1057/978-1-137-56595-2

Library of Congress Control Number: 2016948766

Cover illustration: © Tetra Images / Alamy Stock Photo

Printed on acid-free paper

This Palgrave Macmillan imprint is published by Springer Nature
The registered company is Macmillan Publishers Ltd. London

For my parents, Hilde and Mike Ettrick

ACKNOWLEDGMENTS

My interest in the history of cancer began almost fifteen years ago when I was fortunate enough to be included in the Centre for the History of Science, Technology and Medicine (CHSTM) 'Constructing Cancers' project (supported by the Wellcome Trust Programme Grant 068397) led by the late (and very much missed) John Pickstone at the University of Manchester. Members of that original group—especially Carsten Timmermann, Elizabeth Toon, Emm Barnes Johnstone (now of Queen Mary, University of London), and of course John before his untimely death—proved to be great colleagues over the years and I am extremely indebted to all of them for helping me to bring this book together. Thanks too go to the organizers and participants of the project-related workshop 'How Cancer Changed: Expanding the Boundaries of Medical Interventions' held in Paris at the Centre de Recherche Médecine, Sciences, Santé et Société (CERMES) in 2009. Jean-Paul Gaudillière and Ilana Löwy were particularly helpful with their detailed feedback on the paper that I presented there, a paper that would go on to form the basis of the final three chapters of this book.

The Palgrave Macmillan editorial staff were a pleasure to work with and my thanks go to Jade Moulds, Jen McCall, and Peter Cary for their patience and assistance as I slowly got my act together and completed the book. I am grateful for the comments of an anonymous reviewer as well as those of Carsten Timmermann as he read of the final draft of the book. My former colleague at the University of Leeds, Adrian Wilson, was also very helpful as I wrote the early chapters of the book, and it is down to him, and his long mentorship and friendship, that my interests in the history of cancer extend back beyond the 1880s at all. On this side of the Atlantic the support of the members of the Center for Public History (CPH) at the University of Houston have been invaluable. It is through the CPH that I had the opportunity to circulate and present earlier drafts of this book. My particular thanks go to the remarkable head of the CPH, Marty Melosi, as well as to Julie Cohn and Jimmy Schafer for their many smart and extremely helpful criticisms and suggestions for improvements. My philologist colleague Richard Armstrong was kind enough to oversee my attempts to follow the progression of ideas conveyed by some ancient Greek and Latin medical terms, but here as elsewhere in the book any errors and inaccuracies are mine and mine alone.

Thanks must also go to my academic home since 2005, The Honors College at the University of Houston. It is here with the support and guidance of my Dean, Bill Monroe, that I have been able to translate a (very Pickstonian) passion for the role of the humanities in public life into the creation of a programme in Medicine & Society to serve our pre-medical and other prehealth students in collaboration with local medical and nursing schools. Working within a large urban state university with a mission to serve the sons and daughters of such a culturally, ethnically and economically diverse city as Houston, Texas has presented interesting challenges and enormous opportunities to do good things, and I feel privileged to have participated in the growth of the institution over the past decade. Almost everything in this book has been discussed in Honors classes with undergraduate students at one time or another, and I owe a debt of gratitude to them for the insights I gained as they responded to my ideas with their usual mix of gusto and good-humoured scepticism.

On a personal note, I would like to thank my father's physicians Dr Bernard Leclair of Bazas, and Dr Nathalie Bonichon of Bordeaux, for caring for my whole family in the period following papa's prostate cancer diagnosis. My friends and colleagues Julie Anderson, Bill Monroe, Aaron Reynolds, and Brenda Rhoden provided much needed advice and encouragement during some bleak periods in the writing process. I am grateful too for the support of my family and friends in France, the UK, and in Texas for putting up with being ignored for months on end, and for the love and steadfast companionship of my wife, Carrie. Final thanks go to Flowers the cat for not being afraid to decide for me when it was time to take a break.

Contents

CHAPTER 1

Introduction: The Prostate, Cancer, and the Making of Modern Medicine

How can the very old come to define the very new? The ailments that make up a collection of diseases labelled 'cancer' are described in ancient manuscripts, depicted in millennia of human artifice and exposed within prehistoric human remains. As a species we have always lived with malignant tumours and wasting death. Nevertheless, there is something undeniably *modern* about cancer.[1] For over a century, the control of cancer has perhaps been the ultimate test of our medical prowess, a yardstick measuring our incremental control over nature and a testament to our unwavering expectation of longer, healthier lives, unhampered by disease and disability. The capricious and intractable nature of cancer has not, by and large, done much to sink our buoyant confidence in scientific progress but it has introduced a paradox, widely felt if not always acknowledged, that all is not well in our scientific age. The history of cancer in the twentieth century is at one and the same time a story of extraordinary optimism for a future mediated and enhanced through technology and a story of human fear and frailty in the confrontation of nature and technology. Charles Rosenberg described his view of this paradox of modern medicine in his book, *Our Present Complaint*, saying that we have,

> a characteristic disconnect: on the one hand, uncritical faith in the power of the laboratory and the market, on the other a failure to anticipate and respond to the human implications of technical and institutional innovation. And one of those dilemmas grows directly out of our expansive faith in

© The Editor(s) (if applicable) and The Author(s) 2016
H. Valier, *A History of Prostate Cancer*,
DOI 10.1057/978-1-137-56595-2_1

technological solutions to clinical problems; as we are well aware, sickness, pain, disability, and death are not always amenable to clinical intervention. In the late twentieth century, such conflicts are both public policy issues and—inevitably—elements in individual physician–patient relationships.[2]

Understanding and articulating this 'disconnect' as Rosenberg describes it is at the heart of this book. How *did* cancer come to represent our greatest hopes *and* our most cynical fears for and about the biomedical enterprise?

In writing this book I have chosen to focus on just one cancer—prostate cancer—for a number of reasons, but primarily because it is a very common cancer with little said of it by historians and social scientists and one that perfectly exemplifies the paradox described above. The overwhelming focus of the existing historical literature on cancer has been on breast cancer and while this has been in many ways extremely worthwhile in exposing issues of gender inequality, medical and political paternalism, and issues of activism and so on, it does rather beg the question of why prostate cancer is so under-researched. The two cancers are after all in many ways strongly analogous if we consider what they have to say about social, cultural, and medical interpretations of gender, sexuality, and aging. It is my hope that other researchers with interests in these topics might subject prostate cancer to the same kind of detailed, rigorous analysis that has provided breast cancer and breast cancer patients with such a rich social and cultural history. It is not my intention in this book, however, to write a male version of the existing breast cancer literature. The history of prostate cancer has much to offer on its own account—from a sexualized and pathologized account of masculinity appearing in the new scientific age, through to the creation of new spaces in academic medicine after WWII with integration of the (overwhelmingly male) patients of the Veterans Administration (VA), and the rise of activism that interpreted prostate cancer as part of a systematic exclusion of the interests of men and the male patient from mainstream medical attention—this book covers ground only patchily dealt with by existing literature, and, as such, I hope this book with serve as a meaningful contribution to the literature on the history of cancer. To take just one example, the recent controversy over the use of prostate-specific antigen (PSA) testing as a screening tool reveals so much of what is at the heart of Rosenberg's 'complaint'—particularly as it concerns overdiagnosis and overtreatment—and yet that phenomenon too has received little attention from historians and social scientists.

My focus on academic elites in this book leaves it open to (not unreasonable) accusations that it is a kind of 'great man' history of medicine. The many remarkable studies I discovered while working on this project *have* caused me to single out the brilliant work of several individual researchers. In all the ways that matter though, this is not, I think, a hagiography or any kind of history of that narrow type. As I try to make clear throughout the book, the researchers did not make their famous discoveries as feats of virtuosity so much as they were the end results of collaboration between many men and, of course, women, whose work in the wards, clinics, and laboratories made transformational work practicable. That is simply the way science operates, especially as it became more complex in the long twentieth century. As I also try to make clear, the institutional frameworks in which these researchers operated—whether in the availability of careers, funds, space, equipment, or patients—are crucial context. The final part of this brief *mea culpa* such as it is concerns the patient and his lack of voice in this book. This is a regrettable absence, and one I hope that this account by providing a resource for future historical studies on prostate cancer might help to ameliorate. To this end I have, when appropriate, delved into the political, economic, and cultural life of the disease, but there is much to be done if we are to have a history of male cancer as rich and instructive as that for breast cancer in women.

It might seem sensible to have started this study in the nineteenth century when prostate cancer was for the first time becoming widely discussed and debated in the newly forming era of scientific medicine. I decided to go further back than that in an attempt to do some justice to a story as old as humanity—the terrible sufferings of men unable to pass their urine and the efforts of healers who tried to help them. As I describe in Chap. 2, sympathetic and compassionate accounts of these miseries date back thousands of years. That men experienced this painful, life threatening, 'strangury' as a consequence as of their aging was well known to the ancient healers with education enough to record their practices (and more than likely to the many who hadn't and didn't). Doubtless, much of what they described we would now consider to be benign prostatic hypertrophy (another condition ripe for historical analysis), but such was not understood until much later. I have written inclusively in these early chapters of about 'prostatic enlargement', understanding that causes other than cancer were at the root of the symptoms recorded in the annals of medicine.

We can see in the palaeoarchaeological record that cancer has been with us throughout our history but what we mean by the term 'cancer' has shifted and changed in often confounding ways. The word itself is a Latinized form of the Greek word *karkinos* found in the writings of the Hippocratics, but we might also reasonably claim that the idea of cancer is a much newer phenomenon than that arising from the cellular vision of the body and disease worked out by Rudolf Virchow and his colleagues during the mid-to-late nineteenth century.[3] If I included in this book everything 'cancer' signified in the Hippocratic sense, I would have to write a history of inflammation, a treatise on the soft and hard tumours, and an account of venereal disease, to name but a few things.[4] It is worth the effort, though, I think, to feel back in time and to not just pick up the story on the more familiar ground of ground of nineteenth century laboratory sciences.

Chap. 2 is also a story about anatomy and the changing nature of learned medicine. As the new spirit of autopsy (from the Greek *autopsia*, to see for oneself) permeated the dissecting halls of the great medical schools of the European Scientific Revolution so we get, thanks to Andreas Vesalius in the sixteenth century, the first detailed description of the prostate as an organ involved in reproduction. In the eighteenth century the anatomist Giovanni Battista Morgagni turned anatomy to the study of diseases, looking to locate and analyse lesions in the postmortem body that corresponded with symptoms in life.[5] Morgagni also recognized the prostate and regarded it as an important seat of disease, something likewise taken up by the famous eighteenth century surgeon John Hunter.
[6] Old boundaries between physicians and surgeons were breaking down by Hunter's time, and I use his work on the prostate to examine just why and how that was happening. Chap. 2 concludes with a review of 'cancer' as the concept was understood by the mid-to-late nineteenth century, both by laboratory scientists like Rudolf Virchow and by clinicians observing cancer, particularly prostate cancer, in their practice.

Chap. 3 is a study of how issues of cancer and diseases of the prostate were linked to the growth of urology as a surgical specialty. Ancient techniques to relieve urinary problems in men survived relatively intact well into the eighteenth and nineteenth centuries. What had changed a great deal more than the old instruments and practices of surgery by this time though was how diseases treated surgically were coming to be understood and investigated. Once again John Hunter appears in this account because it was he who did so much to place urology on a learned, academic footing particularly with his work on comparative anatomy. Although he himself stopped short of recommending it, Hunter's observations on the role of the testicles in the function of the prostate encouraged some surgeons to try to use castration as a means of controlling prostatic disease. These operations were highly controversial and it is instructive to look back on the terms and tone of the debates especially as they coincided with moves to craft urology as a recognized surgical specialty at the turn of the nineteenth century.

Although not by any means uniquely American, the push to specialization in the US was particularly rapid as large organizations, including hospitals and universities, looked favourably on the philosophies of scientific management coming out of industry and brought them to their own institutions looking to increase efficiency and increase productivity.[7] One of this group of new specialists was the surgeon Hugh Young whose hugely influential work at Johns Hopkins in the early part of the twentieth century did a great deal to raise the profile of urology even as other surgeons despaired of ever emulating his successes. Young aside, there was an air of gloom within urology during the 1910s and 1920s. By then specialists had become adept at diagnosing prostate cancer even as they were quite fatalistic about what they could then do about it. Some perceived this issue to be one of timing: if general practitioners could be taught to not delay referring patients then they might have more of a chance to intervene. Others still believed that they were doing good by intervening surgically even in advanced cancers and once again we see how debates about restraint and heroic intervention can reveal much about specialties in the making.

Chap. 4 opens with a discussion of the new scientific experimentalism of the mid-to-late nineteenth century, exemplified by the research and writings of Claude Bernard. Along with bacteriology, immunology, and pharmacology, experimental physiology was one of the laboratory sciences underpinning a new style of 'biomedicine' that helped forge a new identity for academic medicine and by extension to professional medicine as a whole. Abraham Flexner's famous report on the state of North American medical education published in 1910 is usually regarded as the turning point in the professionalization of US medical education, but reformers were certainly very active well before then.[8] Decades before Flexner took his tour of the nation's medical schools to collect material for his report, elite physicians had seen reform of medical education as a means to regulate the profession as a whole by tightening and restricting the route into licensed practice. Indeed, Flexner himself made good use of these reform-minded elites when he held up Johns Hopkins School of Medicine—a school itself modelled on the German academic medical system—as an ideal and a model to be emulated elsewhere. The Flexner report does, though, act as my turning point in this book. The US-focus that began in Chap. 3 continues for the remainder of Chap. 4 and is the exclusive emphasis in the chapters that follow. There is much to be said about prostate cancer beyond the US, of course, and I hope that others will say it. Because the historical literature on this common cancer *is* so small, though, the US demands the attention I give it in this book because of the sheer volume of important work that was done there. The elucidation of the biological nature of prostate cancer and the development of the means to treat and detect it is an overwhelmingly American story.

History is seldom about the new replacing the old,[9] and this is beautifully shown by what happened when the brilliant prodigy of Bernard, Charles-Édouard Brown-Séquard, revealed his glandular theories (and glandular extracts) to the world.[10] Embedded within and emerging from the experimental physiology that academic elites celebrated for the intellectual and cultural capital that it brought to them, Brown-Séquard's work nonetheless found a comfortable place in the 'old-style' medical marketplace in the US. The obvious titillations of testicular extracts and the 'masculine rejuvenation' they promised brought out some of the best (or at least notorious) of that old style, such as the great showman, John R. Brinkley (known across the country as the 'goat-gland doctor'). While organizations like the American Medical Association (AMA) despised and despaired of such charlatanism, testicular extracts show that the old and the new styles of medicine existed cheek-by-jowl well into the 1930s. This was not simply a case of orthodoxy *versus* quackery, however. The quasi-respectable provenance of glandular theories (Brown-Séquard had an impeccable scientific pedigree but he certainly attracted criticism) caused lines of respectability to become blurred. This uneasiness continued as several 'rejuvenating surgeries' came into vogue—vasectomies, testicular implants, and the like. The chapter shows that the American medical marketplace was stubbornly pluralistic well into the interwar period of the twentieth century, something that we would do well to remember when thinking about the rise of the 'patient-consumer' as a phenomenon dating to the (highly politicized) period of 1970s and 1980s medicine.[11]

It perhaps not surprising given this context that early attempts to educate men about problems with their prostate and 'intimate health'—urinary and sexual—were heavily moral in tone. Self-help texts in the early part of the twentieth century encouraged men to seek out practitioners of the (still nascent) specialty of urology as a routine part of their self-care as they aged. In spite of the often overtly censorious way in which the aetiology of prostate problems was discussed (especially with respect to masturbation), such texts tended to be written with the aim of reducing the shame and secrecy that might prevent men from seeking out appropriate medical advice. Professional interests were also at stake in these efforts—men overcome with embarrassment might well avoid the medical encounter entirely and instead seek out the snake oil salesmen, something that was anathema to a profession so concerned to protect its image and status.

What really transformed the relationship between patients and specialists was another blossoming in the alliance between the laboratory and the bedside, albeit one grounded in the theories of hormones and glands. When a young Canadian physician, Charles Huggins, arrived at the University of Chicago in the 1920s to pursue his interests in experimental physiology, he was part of an as yet still small cadre of academic clinician researchers in the US. At Chicago, Huggins enjoyed purpose-built facilities that brought together laboratories for animal experiments, clinical chemistry (biochemical) laboratories for analysis, and wards supplying 'clinical material' (patients). Huggins would make extensive use of all three kinds of resources in his (Nobel Prize winning) discoveries of the late 1930s and early 1940s that made clear that many kinds of prostate carcinomas depended on hormones for their growth, and that these cancers could, furthermore, be damaged if their hormonal nourishment was disrupted or 'ablated'.

Huggins and his team treated patients, some with very advanced cancer, with a synthetic hormone called diethylstilbestrol (DES) that acted against the testosterones they understood to be fuelling the growth of carcinomas. While often not curative, the therapeutic use of DES resulted in many remarkable changes in how men with prostate cancer experienced their disease. DES was often used to alleviate some of the more painful and troublesome symptoms experienced by men with advanced cancer, and, for many, it changed the course of their disease from an acute to a chronic course. Until Huggins' hormone therapy came into widespread use, urology was almost an entirely instrument-based surgical discipline, and one whose practitioners, while able to intervene in a wide variety of urinary problems, could still do little for patients with symptomatic prostate cancer. Following Huggins, urologists and academic researchers gained a powerful therapeutic tool and an entirely new research strategy. Huggins' obsession with finding methods to quantify his monitoring of his experimental patients also helped to make hormone-based therapy and research viable. While he was not the first scientist to describe phosphatases, his extensive work on prostatic acid phosphatase (PAP) in the 1940s represented a profound shift in how men with prostate cancer could be diagnosed and treated. PAP levels were often very elevated in men with prostate cancer, but it was recognized that elevation alone was not a reliable enough diagnostic indicator to be used in isolation. What PAP did do, though, was allow clinicians to better understand what the prostate was doing as the patient's carcinoma was treated with DES. Indeed, PAP remained a central part of the way clinicians planned and monitored the treatment of prostate cancer patients until its use as a biological marker was eclipsed by PSA testing in the late 1980s.

The ability to 'see' the prostate via a biochemical marker although incredibly useful did not *directly* reveal much about the cancer itself. In Chap. 5 I describe how work in the 1950s and 1960s by a pathologist named Donald Gleason helped to change this, and in doing so opened up yet more opportunities in diagnosis, treatment planning, and research. Like Huggins, Gleason was very much both a liminal figure *and* a product of his time when he did his most famous work: the creation of a pathological grading system for prostatic carcinomas that still bears his name, the so-called the 'Gleason score'. I say 'liminal' because Gleason worked in the federal Veterans Administration hospital system (later Veterans Affairs, VA), which was in the immediate post-WWII years a fairly inauspicious place to be. Gleason was also very much 'of his time', however, because of the very modern way he established his discoveries as valid and reproducible (critical components of any scientific breakthrough) within the new investigational tool of the multicentre, cooperative, randomized clinical trial (RCT). The VA, by virtue of its clientele, had an abundance of patients with prostate issues and this, too, served the interests of the new clinical trial enthusiasts. Histories of clinical trials in cancer are not numerous but where they do exist they tend to focus on chemotherapies, particularly as applied to juvenile populations.[12] By contrast, the VA trials cancer trials focused on common and hard to treat cancers (like prostate and lung) and represented a quite different kind of endeavour that is as yet unwritten but which is nevertheless crucial if we are to understand why and how such a core component of biomedicine as the cooperative clinical trial developed in the ways that it did.

Legislation enacted during the war intended to brace the VA for the strains it would face as veterans flooded back from the war also contained within it a new means to organize and commission large-scale clinical investigations within the system. Although the relationship between academic medicine and VA would grow to be intimate, in the decade or so after WWII it was not obvious that this would be the case. In the late 1940s medical schools and teaching hospitals were called upon to provide clinical services and support to the VA. What helped to cement the relationship and produce a permanent intertwining of the VA and academic medical communities, however, lay not so much in what academic medicine could do for the VA as with what the VA could do for academic medicine. An important part of this reciprocal relationship began with the Servicemen's Readjustment Act of 1944, or, as it's more commonly known, the GI Bill. Large numbers of demobilized veterans seized the opportunity to be funded in their education and training in the professions. Medicine was an attractive route into this professional class, but a surge in medical students would create problems—in this instance the needs of veterans would strain not the VA but rather the resources of the medical schools and their associated teaching hospitals. Here the VA provided valuable capacity, both in terms of overall patient numbers but also in the tractability of patients as regards the needs of clinical instruction.

As the ways in which US patients paid for healthcare changed in the 1950s and 1960s, especially with the rise of employer-sponsored private health insurance, so patients in VA hospitals and clinics—a system designed to be free at the point of access—differed from the well-insured, fee-generating patients then being courted by the academic medical centres. Teaching hospitals after WWII actively sought such patients as a means to offset the financially punitive provision of medical education and, as was often the case, care for the medically indigent for which they were unlikely to receive anything more than minimal reimbursement. In the first book to be published as part of his autobiographical series of reflections on the state of modern medicine, Atul Gawande's *Complications: A Surgeon's Notes on an Imperfect Science* (2002), describes how the burden of training new generations of doctors has not been equally borne by patients. Although not necessarily by design, well-insured, fee-paying patients have, nonetheless, been much more likely to avoid the clumsy advances of the inexperienced clinician than other categories of patient: a situation that Gawande acknowledges as understandable even as it is, to his mind, morally objectionable.[13]

The role of VA patients as clinical material in academic medical research is a history that again remains in large part unexamined by historians. There are numerous fine studies on the troubling record of the US military's involvement in secretive human subject research during WWII and the Cold War,[14] but there is much less on how veterans were enrolled and treated in academic clinical trials. In the late 1990s a very public scandal erupted that temporarily shut down *all* research at the VA after clinical investigators at the VA's West Los Angeles Healthcare Center were reported for failures to gain informed consent from research participants.[15] The investigation that followed revealed decades of problems and discontent all of which could (and should) be usefully explored by the interested historian. My intention here though is not to add to the literature of scepticism and criticism of the ethics of clinical trials, as worthwhile as those studies are. The vast majority of clinical trials have, like the studies described in Chap. 5, enjoyed an unblemished reputation through a strict adherence to the new rules at the bedside—rules that came to be known as 'bioethics'.[16] In some part due to the uncovering of scandals in human subject research, and in some part due to the new ethical dilemmas posed by novel technologies (for instance, who should receive dialysis while others died? or when could life be said to be extinct in a patient supported via mechanical ventilation?) academic medicine responded to forces within and without to participate in a more transparent, if more chaotic, ethical age. Instead of any ethical critique then, what I do hope to do in this chapter is add to a literature seeking to understand how and why clinical trials became such an integral part of how we investigate the many diseases grouped together under the umbrella of 'cancer'.

In Chap. 6 I move on to consider what has been, in many ways, *the* defining aspect of prostate cancer at the end of the twentieth century—the arrival of the PSA test in the 1980s. The PSA test literally and figuratively brought prostate cancer into the public eye as manufacturers of the test launched ambitious 'public awareness' campaigns fronted by high profile celebrities encouraging men to take care of their health by taking the test. As had been the case for breast cancer activism before it, prostate cancer activists (often directly assisted by those who had so successfully pushed the agenda for breast cancer[17]) seized upon the spirit of patient-as-consumer and joined with industry and academia to promote the cause and to lobby for state and federal support of prostate cancer. In spite of concerns early on that the PSA test was a measure of *prostate* function not *cancer* function, the test became wildly popular—sought out by patients and enthusiastically promoted by specialists backed by recommendations from the major urological professional associations.

As a direct consequence of all of this new testing, the incidence of prostate cancer amongst men in the US underwent an enormous surge— the so-called prostate cancer 'epidemic' of the 1990s. For patients in much of the rest of the world elevated levels of PSA might mean entering a period of 'watchful waiting', but in the US elevated PSA levels tended to mean rapid intervention. Rates of prostatectomy skyrocketed as surgeons urged their patients to 'get ahead' of the disease and to remove it while it was still encapsulated within the prostate. Critics protested strongly to all of this, pointing out that many men had *some* malignancy of the prostate (which may or may not ever rise to become a clinical cancer), so what biopsies went looking for they inevitably found. As prostate cancer became embroiled in wider controversies over cancer screening[18]—controversies that had mostly been generated by accusations that too many unnecessary interventions in the breast were occurring as a result of routine mammograms—two major RCTs reported on their more than decade-long studies of the usefulness of PSA screening in saving lives. The results did not make good reading for advocates, nor did the 2012 report by the US Preventative Services Task Force that graded PSA screening a 'D' grade (meaning not recommended due to harms outweighing benefits, in contrast an earlier Task Force had awarded mammography a 'C' grade—not encouraged except in selected cases where individual circumstances might warrant it). The backlash against the recommendations from all quarters—activists, clinicians, industry, professional organizations—was ferocious. Researchers were accused of many things including being in cahoots with an administration then championing a piece of healthcare legislation known as the Affordable Care Act (or more colloquially, Obamacare), which critics argued was a form of 'socialized medicine' that would bring in healthcare rationing and reduce patient choice.

It's also interesting to see how critics evoked problems with the methodology of the Task Force recommendations, based as it was on 'evidence-based medicine' (EBM). Clinicians and specialist groups raged that EBM, which depended on a wide reaching literature and statistical review (meta-analysis) of years of reporting about trials and studies of particular interventions, had little relevance to 'real life' practice. Objections to the principles of EBM and RCTs are further discussed in Chap. 7 where I shift my study away from hormones and surgery and consider the role that radiotherapy has played in the treatment of prostate cancer. Aside from some early experiments on brachytherapy (the insertion of radioactive matter into or adjacent to the prostate) and some attempts to use radioactivity as a means to 'castrate' the patient (by killing off testosterone production) not much radiotherapy was tried for prostate cancer. The reason for this has a great deal to do with the anatomical position of the prostate—deep as it is inside the bony pelvis. The kinds of exposure necessary to penetrate that deeply would cause massive damage to the skin and intervening tissues, so it was seldom attempted. With the introduction of linear accelerators and the first cobalt-60 machines after WWII came high energy x-rays and gamma rays, respectively, and so new opportunities for external beam radiotherapy to penetrate more deeply into the body. Aside from these experiments with high-energy photons though, research into the medical applications of fast particle beams was becoming possible as cyclotrons (and later synchrotrons) built in physics laboratories from the 1930s onwards became more sophisticated. Although not designed for patient care, particle beam therapies were tried out on various kinds of cancer with local hospitals providing patients brought in between physics experiments, to decidedly mixed results.

By the mid-1950s some hospitals in the US and in Europe had started the process of acquiring their own purpose-built cyclotrons for the purposes of neutron therapy. Within a decade, however, it was becoming obvious to that the huge investments in neutron therapy as an anticancer modality were simply not paying off. Still, the disappointments of neutron therapy did not mean the death of fast particle research, far from it: attention turned instead to the production of proton beams. By 1990, the first purpose-built proton therapy facility was opened in Loma Linda University Medical Center in California (soon followed by others across the country). Again, the proton therapy story is not a *uniquely* American one, but it has a particular American flavour. The speed with which the technology evolved as an anticancer modality in the US was beyond compare. As of 2015, the US has more proton therapy machines than any other country by a wide margin (there are two just in the state of Texas, for instance), so the question becomes why? The answer, I think, has a great deal to do with the issues discussed in Chap. 6, specifically in the ubiquity of the PSA test for prostate cancer. As more and more men were being tested so more and more 'early-stage' cancers were detected (that is, those encapsulated in the prostate), the many boosters of proton therapy marketed the technology very aggressively as the ideal way to treat this patient group.

This claim to efficacy in proton therapy hinges on rationales drawn from physics and pathophysiology, namely that it is a quality of particles (unlike photons) to release their energy at their stopping point meaning that tissues before, after, and surrounding a tightly focused beam would not be affected in any way deleteriously. Proton beams therapy was, in other words, all about *precision*. The problem of this for critics was that none of this had actually been *shown* to be the case in RCTs. Advocates pushed back, basing their arguments on appeals to ethics and practicality. They argued that since protons were *clearly* better, a trial in which patients were placed in 'inferior' treatment arms was unethical (an argument known as clinical equipoise). They also pointed to the recent history of radiotherapy, which had seen both conformal radiotherapy and intensity-modulated radiotherapy (IMRT) introduced without trials. EMB reviews of what clinical studies did exist, including a preliminary release from the Agency for Healthcare Research & Quality (the same organization that oversees the Task Force committees), were not encouraging about the benefits of proton therapy, at least in relation to prostate cancer. This presented serious problems to operators that depended on a large throughput of patients to provide a return on investments for these hugely expensive facilities. With prostate cancer so commonly diagnosed (especially in the States) these patients provided a large pool of potential consumers. Such was not the case in other types of cancer that *were* more positively indicated for proton therapy, such as ocular and paediatric cancers, as they affected so many fewer people overall. The chapter closes then with a consideration of whether the multi-billion dollar 'proton bubble' is about to burst.

Chap. 8 is in the form of a conclusion, and I bring together some loose ends and speculate on what the ongoing experience of prostate cancer is likely to be for those men diagnosed with it. I also consider how prostate cancer came to be viewed as a 'neglected' disease (an idea that I criticize), and what the consequences of a perceived 'marginalization' have meant for gendered messaging around masculinity, health, and the identity of the male patient. My final thoughts are to consider, by way of a personal reflection, how we might all do a little better in the face of the paradox at the heart of Rosenberg's 'present complaint'.

NOTES

1. Historians generally use the phrase 'modern medicine' to refer to that period from around the 1500s onwards to describe the trends in post-medieval European medicine: more hospital building (religious and civic), ongoing formation of professional organizations and competition, more systematized research, newer techniques and therapeutics to name a few, but here I am using the word 'modern' to mean both 'recent' and to evoke a sense of modernism in medicine, a radical reconceptualization of the possibilities and limitations of technologies, organizations, and the new social and political environment of medicine in the second half of the twentieth century.

2. Rosenberg, *Our Present Complaint*, 169.

3. Rather, *The Genesis of Cancer*.

4. This is an idea I take from Lawrence, Medicine The Genesis of Cancer.

5. An impressively comprehensive account of anatomy as practised around the time of Morgagni appears in Cunningham, *The Anatomist Anatomis'd*.

6. For an account of the life and times of John Hunter, see Moore, *The Knife Man*.

7. An exhaustive account of specialization in this period can be found in Weisz, *Divide and Conquer*.

8. For a useful discussion of this view, see Barr, Revolution or Evolution?

9. Like much of the rest of the analysis in this book, my comments here reflect the work of John Pickstone, see in particular Pickstone, Ways of Knowing, 435.

10. A book-length study of the rise of glandular medicine is provided by Sengoopta, *The Most Secret Quintessence of Life*.

11. For one well-known example of the politicization of patienthood in the 1970s, see Boston Women's Health Book Collective, *Our Bodies Ourselves*.

12. See Keating and Cambrosio, *Cancer on Trial*, for a comprehensive study of this type.

13. Gawande, *Complications*, 11–35.

14. See, for example, Goodman, McElligott, and Marks, *Useful Bodies*.

15. Hilts, V.A. Hospital Is Told to Halt All Research.

16. On the emergence of bioethics, see Rothman, *Strangers at the Bedside*.

17. For a thought-provoking study of breast cancer activism, see King, *Pink Ribbons, Inc.*

18. For an excellent and concise overview of this controversy, see Rosenberg, Managed Fear.

CHAPTER 2

The Problematic Prehistory of Prostate Cancer

*A sharpe humour which flows from the glandules termed prostatae and con-
tinually runs alongst the urenary passage, in some places by the way it frets
and exulcerates by the acrimony the urethra in men. In these, there sometimes
growes up a superfluous flesh, which oft times hinders the casting or coming
forth of the seed and urine by their common and appropriate passage, whence
many mischieves arise.* Ambroise Paré, *Des Chaude-Pisses, des Pierres et des
Retentions d' Urine* (1564)[1]

In the early summer of 1817, the Professor of Anatomy and Surgery
to the Royal College of Surgeons, William Lawrence, presented several
cases of 'Fungus Haematodes' (including some reported by his colleague, George Langstaff, the attending surgeon for St Giles' Cripplegate
Workhouse). As part of his series, Langstaff recorded a case of extreme
urinary blockage arising from, he supposed, a diseased prostate:

> I.B. a pauper, sixty-eight years of age, had laboured under an affection of the
> bladder upwards of five years, and had been under the care of several sur-
> geons without experiencing any essential relief. During the last six months
> of his life, he had suffered the most excruciating pain in the region of the
> kidney and bladder, attended with almost constant desire to void urine,
> which was effected with the greatest of difficulty, either by drops, or in a
> very small stream, and generally coloured with blood. … An examination
> per rectum, proved that there existed an enlarged state of the prostate gland,

© The Editor(s) (if applicable) and The Author(s) 2016
H. Valier, *A History of Prostate Cancer*,
DOI 10.1057/978-1-137-56595-2_2

and slight pressure occasioned great pain. A bougie of moderate size was introduced into the urethra to ascertain the state of the canal. It passed readily as far as the membranous part, but could not be conveyed beyond it; and it was with the greatest difficulty one of the smallest kind was made to enter the cavity of the bladder.[2]

Noting that the final, feverish decline of his patient was bad as anything seen in the last stages of (the notoriously miserable) typhus, Langstaff records death a mere four days after the first consultation. As an avid practitioner of the new science of 'morbid anatomy' he then went on to describe the subsequent postmortem dissection:

The bladder and urethra were ... examined. The former felt like a solid substance: on laying it open, it was found to contain a tumor as big as a large orange, the surface of which was covered with recently coagulated arterial looking blood ... After minutely examining the tumor, it was discovered to derive its origin from the prostate gland, chiefly from the middle, or third lobe. ... A perpendicular section was made into the tumor, which was composed principally of loose coagulated blood mixed with a white pulpy substance; but its base was of a dense, hardish consistence ... The fungus extended laterally, and had completely plugged up both ureters ... [t]he prostatic urethra was nearly closed with the same growth...[3]

Here we see a prime example of an expert scientific examination of a patient in the early nineteenth century, but one that simultaneously combines the old and the new. Old are the biographical case histories of the patient of the sort collected by the followers of the Hippocratic tradition, and old too is the use of 'bougies' or cylinders inserted into the penis to relieve urinary blockage, which date back to at least ancient Egyptian times, as does the practice of recording observations of the urine collected. What is new here is the combination of these traditional practices into a system of investigation that made use of physical examination and, significantly, a careful correlation between the illness as experienced by the unfortunate patient in life, and the appearance of disease in the parts of the body revealed after death—also known as morbid anatomy. As chief surgeon to a workhouse infirmary, Langstaff had ample opportunity to see a high volume of patients and to dissect their remains: each a necessary condition of practising the new science.

Twenty-six years after the Lawrence publication, we see another case of terminal prostatic disease appear in the literature, this time authored by the London surgeon John Adams. Like Lawrence and Langstaff, Adams emphasized how careful observation of the structural abnormalities seen after death might benefit an understanding of disease in life. Speaking in front of an audience of members of the Royal Medical and Chirurgical Society Adams presented his paper, 'A Case of Scirrhus of the Prostate Gland', about a case he had seen as part of his practice as surgeon to the London Hospital, which, like Cripplegate, was another large urban hospital established to treat the sick poor:

> A gentleman, aged fifty-nine, was suddenly seized with paralytic symptoms, which seemed to arise from the derangement of circulation. During his recovery he experienced a frequent desire to pass urine, and required the constant use of a catheter. The instrument passed over a hard and rough surface, and induration and enlargement of the prostate were felt upon examination per rectum.[4]

In spite of Adams' intervention with the catheter, the patient died three days after admission. In keeping with the times, Adams then dissected his patient and recorded his findings:

> The prostate gland was enlarged to nearly twice its natural size; an ovoid mass, *distinctly scirrhous* [sic], the size of a small nut, projected into the bladder from its upper surface. The left lobe was occupied by a long scirrhous mass; the right lobe appeared healthy.[5] [original emphasis]

In the discussion following the presentation of the case, James Copland the eminent Scottish physician and then President of the Society, noted that in his view scirrhus of the prostate was a very rare disease and asked his colleagues for their comments. He received a mixed response. Some disagreed that scirrhus was all that uncommon while some others agreed that it was but that fungus hematodes weren't. When pressed on the nature of the tumour, Adams was at pains to point out that his specimen had 'been examined by an experienced microscopist, who had pronounced it to be true scirrhus in every particular'.[6] To the gross observations of tumour morphology as detailed by both Langstaff and Lawrence before him then, Adams saw fit to add the opinion of a pathologist trained in the use of the microscope both in order to more perfectly know the structural nature of the tumour and, presumably, as a means to more accurately identify the nosological character of the disease suffered by his patient.

So what are we to make of these two cases, both appearing to describe cancer originating in or at least involving the prostate, but whose authors do not mention the term 'cancer' at all? One approach we might take would be to dismiss their observations and surmises as half-right guesses shrouded in ignorance of the 'true' nature of disease. As I discuss in the Introduction, for many casual observers of the progress of science such an approach is usually ample even though it leaves out from the story precisely that which is of most interest. It is clear that our authors far from being virtuosities regarded themselves as part of a larger community of investigators suggesting and contesting the facts of disease. To pick out these two papers as in any sense 'firsts' in the history of prostate cancer is a fairly arbitrary exercise, but it is an informative one if it leads us to consider some deeper, underlying conditions of the science of the times. To understand these cases in their wider context then, we need to ask a number of questions, including how the prostate and cancer have each been understood historically, and why our authors depended so heavily on joining their clinical observations with intricate description of the internal state of their dead patients.

PHYSICIANS, CANCER, AND THE PROSTATE
FROM THE ANCIENT WORLD TO THE RENAISSANCE

The anatomical position of the prostate, deep inside the pelvis and wrapped around the urethra, helped to keep the organ obscured from medical eyes until relatively recently. In part this was also due to the largely non-anatomical nature of western (or any other) medicine until the late Renaissance period. The foundational texts of the western tradition, the Hippocratic *Corpus*—ascribed either to the fifth century BCE physician, Hippocrates, or those he influenced—were principally concerned with explaining the upsets of the body in natural, rather than divine, terms and in devising theories concerning the environmental influence on disease and the means to influence the health of the body by natural interventions. The Hippocratic writings are diffuse and somewhat inconsistent as might be expected of a body of work produced by multiple authors over the period of a century or more. That said, a dominant theme of the works is a view of the body as a microcosm of the macrocosm, an embodiment of the natural world to be understood via analogy to natural phenomena:

The south wind has the same effect on the earth, the sea, rivers, springs, wells and everything that grows and contains moisture. In fact, everything contains moisture in a greater or lesser degree and thus all these things feel the effect of the south wind and become dark instead of bright, warm instead of cold and moist instead of dry. Jars in the house or in the cellars which contain wine or any other liquid are influenced by the south wind and change their appearance. The south wind also makes the sun, moon and stars much dimmer than usual. Seeing that such large bodies are overcome and that the human body is made to feel changes of wind ... it follows that southerly winds relax the brain and make it flabby, relaxing the blood-vessels at the same time.[7]

When the Hippocratic writers used dissection they did so on animals and theorized the existence of similar states in human beings as a continuation of the continuum of analogies in nature. The author of the treatise, *The Sacred Disease*, for instance, writes about demonstrating the causal relationship between winds, the brain, and the sacred disease (epilepsy) by opening the skull of an afflicted goat and showing a brain 'full of fluid and foul smelling', so demonstrating the natural (not sacred) and macrocosmic nature of the disease.[8]

Beyond the social and religious taboos prohibiting dissection, it made a great deal of sense to extrapolate from the natural world to humans since the basis of Hippocratic disease was humoral, a living system of blood, bile, and phlegm to be maintained through diet and exercise and restored to balance—by bleeding or purging, for instance—that itself corresponded to the conditions, environment, and seasons of the world. The principles of humoralism (some of which we see in Paré's account at the opening of this chapter) endured in medical theory and practice through to the nineteenth century, even as the practice of anatomy began to shape medical education from the thirteenth century onwards. Again this makes a certain kind of rational sense: if dynamism defined health and disease, it was not an obvious proposition to look for answers in the stillness of death. (The works of the *Corpus* that did deal with the solid structures of the body commonly worked from those observations that might be allowed from fractures and wounds as a result of accident and war, while the one dissection-oriented outlier, *The Heart*, appears to derive from a significantly later period.[9])

For the Hippocratics, local accumulations of humors could lead to several issues such as *phyma* (swelling), *oidēma* (soft swellings), *sklēra* (hard growths), or *karkinōma* (named for the Greek word *karkinos* or crab for a reason not explained in what fragments remain of the Hippocratic writings but which later authors interpreted variously to refer to the sensation of 'pinching' pain; to the shell-like hardness of some tumours; or the appearance of a highly vascularized penetrative 'claw-like' structures of others). The descriptions of some patients suffering from *karkinōma* resemble what today we might consider a disease course typical of cancer—pain, wasting, terminal outcome—but it was not clear that the Hippocratics saw this course as inevitable in patients with 'the crab'.[10] To understand better the significance of different designations for Hippocratics then, it is first important to again consider the larger premise of the Hippocratic physicians, in this case the avocation to help and not harm the patient. As close observers of the natural world, the Hippocratics had a very clear sense of the limits on the power of humans to intervene in diseases, with the author of *Science of Man* saying about this, 'I would define medicine as the complete removal of the distress of the sick, the alleviation of the more violent diseases and the refusal to undertake to cure cases in which the disease has already won the mastery, knowing that everything is not possible to medicine.'[11] Such rhetoric set apart the Hippocratics from the spiritual healers and others with who they argued and competed, but it also formed the basis of what is probably the most important piece in understanding how these scholars defined their success as physicians, that is as *prognosticators* giving help and comfort where possible but ultimately acting as a guide to the workings of nature:

> It seems highly desirable that a physician should pay much attention to prognosis. If he is able to tell his patients when he visits them not only about their past and present symptoms [gleaned through the study of similar case histories that were scrupulously collected by the Hippocratics], but also to tell them what is going to happen, as well as to fill in the details they have omitted, he will increase his reputation as a medical practitioner and people will have no qualms in putting themselves under his care. Moreover, he will the better be able to effect a cure if he can foretell, from the present symptoms, the future course of the disease.[12]

To see into the workings of the universe and to prove this mastery through prediction is the basis of all science, and the enormous intellectual and cultural capital that such predictions could bring make this aspect of the Hippocratic strategy appear very modern. To return to the *karkinōma*, however, we see the immediate relevance of the Hippocratic classification of inflammations and growths in guiding the action of physicians. In the *Aphorisms* the author gives the following advice, 'It is better not to treat those who have internal cancers since, if treated, they die quickly; but if not treated they last a long time.'[13] (This was an issue that would become a crucial part of the twenty-first century cancer debate, perhaps particularly in the case of prostate cancer.) For the Hippocratics, intervention could not only harm the patient but experience showed that control of the condition was beyond the abilities of the physician and his better option would then be to warn the patient of their fate and to comfort them as best he could.

The school of medicine founded by Alexandrian physicians Erasistratus and Herophilus around 300 BCE was one of the few places in the ancient world where human dissection—and perhaps even vivisection—was performed, if briefly. With the bodies available for dissection scarce (probably recently executed criminals) the investigations were concerned with understanding normal, not morbid or pathological, anatomy. The great city founded by Alexander the Great in 331 BCE was a haven for knowledge of all kinds as he and subsequent leaders sought to build a citadel of learning in the arts, philosophy, mathematics, and medicine to rival any Hellenistic city state and even Athens herself. The work of the Alexandrians survives only in fragments, but the great medical systematizer of the Roman period, Galen of Pergamon, who himself claimed to have dissected human remains, recorded many extracts from the Alexandrian school, including dissection of the 'glandular bodies' of the male urological system:

> The humor produced in those glandular bodies is poured into the urinary passage in the male along with the semen and its uses are to excite to the sexual act, to make coitus pleasurable, and to moisten the urinary passageway ... This is the reason, I suppose, why they do not hesitate to call the passageways arising from these bodies spermatic vessels and indeed Herophilus was the first to call them 'glandular assistants' (*parastatai adenoeides*) since he had called those that grow out from the testes 'varicose assistants' (*parastatai kirsoeides*).[14]

Here then we see one of the earliest mentions of the 'prostate', albeit in reference to structures that we would not consider to be a part of the prostate as we understand it now. Galen was an avid Hippocratic and was otherwise extremely widely read in and educated in philosophy and the writings of different medical sects, something that helped him to create a grand body of knowledge that linked Hippocratic writings with all manner of other ideas, old and new.[15] Regarding the source of swellings and growths as an accumulation of the humors out of balance, Galen also followed the Hippocratics in advising that cancers in the depths of the body should be left alone. He did though expand on the Hippocratic distinctions of inflammation and growths, including elaborating on the earlier 'hard' growths which he referred to as 'scirrhus' (probably synonymous with the Hippocratic *sklēra*, or hardening) and which related in some way to the *karkinōmata* (cancers), a term and concept that would be taken up by the anatomical investigators of the sixteenth century onwards, including our own John Adams and his case of a scirrhus of the prostate. Galen also incorporated post-Hippocratic notions, such as the term *onkos*, or tumour (literally, bulk or mass), that probably originated in the writings of the second century Greek physician Soranus of Ephesus, and that also would go on to be appropriated for a much later generation of physicians specializing in cancer, the 'oncologists' of the late nineteenth century and beyond.

We don't have much evidence of how the Hippocratics dealt with the non-hidden or superficial cancers, but Galen describes several surgical techniques for dealing with tumours, notably for a kind that he saw as one of the most common, schirrus of the breast, that he believed could be effectively treated with excision and cauterization.[16] Whatever the scope of Galen's human dissections, he certainly did create of a substantial body of work relating to dissection and vivisection using monkeys, dogs, pigs, and a variety of other animals. Indeed, Galen mounted some spectacular public displays to demonstrate his technical skill and the veracity of his underlying concepts of the workings of the animal body:

> Once I attended a public gathering where men had met to test the knowledge of physicians. I performed many anatomical demonstrations before the spectators; I made an incision in the abdomen of an ape and exposed its intestines: then I called upon the physicians who were present to replace them back (in position) and to make the necessary abdominal sutures—but none of them dared to do this. We ourselves then treated the ape displaying our skill, manual training, and dexterity. Furthermore, we deliberately severed many large veins, thus allowing the blood to run freely, and called upon the Elders of the physicians to provide treatment, but they had noth-

ing to offer. We then provided treatment, making it clear to the intellectuals who were present that (physicians) who possess skills like mine should be in charge of the wounded.[17]

Such was the incredible rhetorical power of dissection, but the practice, on the human at least, would once again dwindle away as the medical texts of Hippocrates, Galen, and the rest were largely lost to European eyes, enjoying instead a rich existence to the east where they were translated and elaborated as part of a great new age of Arab and Islamic medicine.

The Italian and French universities of late Renaissance were where two great historical confluences met that would reignite the desire to pursue anatomy: the prevalence of ancient texts recovered to the western world (along with their Arabic and Islamic annotations); and the moral and legal permission to open and investigate the human corpse. The Dutch physician Andreas Vesalius was a product of his time in that he was an enthusiastic 'humanist' devoted to the study of the ancient texts better to understand the nature of the world and the place of humans within it. As a professor at the University of Padua he was also able to read the texts of the great medical writers, especially Galen, which were, he first assumed, written from the human body. When he realized that the mistakes of Galen meant that the great man had *not* based his work on human dissection, Vesalius set out to make these observations for himself.[18]

While anatomy professors before Vesalius had instructed students on the works of Galen with use of the body, they had done so in the manner of the first post-Alexandrian anatomist Mondino de Luzzi, who worked in early fourteenth century Bologna, that is, from behind a lectern while assistants carved the body to demonstrate the truth of the texts as students looked on. Vesalius did not lecture in the same way; he encouraged his students to follow in his footsteps by first opening and then 'reading' the body for themselves. It was to be his interest in another of the great Renaissance fashions—that of exquisitely intricate visual representation— however, that changed his life and the world around him forever. Working with his illustrator Jan van Calcar, a student of Titian, Vesalius began the enormous undertaking that would become one of the defining texts of the Scientific Revolution, *De Humani Corporis Fabrica* (The Fabric of the Human Body), published in 1543.

In book V of *De Fabrica*, 'The Organs of Nutrition and Generation', we see a discussion of the prostate, which Vesalius (guided by Galen's descriptions of Herophilus' findings) saw as a glandular body at the neck of the bladder into which inserted the vessels carrying semen from the testes.[19] *De Fabrica* was of huge and immediate influence and was soon incorporated, copied, and annotated into other medical texts, and this is probably how knowledge of the existence of the prostate first began to disseminate to a wider audience of learned physicians. As the account of the famous French battle-surgeon, Paré, quoted at the beginning of this chapter shows, the existence of the gland was certainly known twenty years or so after *De Fabrica* first appeared. Another contemporary reference to the gland is found in the work of the Montpellier physician André du Laurens, *Historia Anatomica Humani Corporis*, first published in 1593.[20] In his treatise, Laurens borrowed heavily from Vesalius' text and illustrations while adding commentary of his own. It was in this commentary that Lauren's recorded the colloquial name for the glandular body beneath the bladder (which he assumed to be in two parts) to be *prostatae*, an observation that the historians Franz Josef Marx and Axel Karenberg consider to be an incorrect rendering of the Latin translation of the Galenic, *parastatai*, or 'assistants'. The closer sounding word, *prostates*, meaning one who stands before, or protector, did not appear in the ancient texts in reference to the gland or any other medical or anatomical feature but the neologism had an obvious metaphoric appeal to later writers looking for 'authentic' Latin roots. Laurens' mistaken use of *prostatae* to reference two glandular bodies at the base of the bladder was the one that gained traction and, as more medical texts came to be printed in the common languages like English rather than in Latin following the introduction of the printing press, the term 'prostate' came into common scholarly use by the early eighteenth century. It is to this century that I now turn.

THE PROSTATE IN THE AGE OF MORBID ANATOMY

The incredible body of anatomical work assembled in Europe during the sixteenth and seventeenth centuries made possible the new science of morbid anatomy in the 1700s. Much of the anatomy conducted to this point was performed on the normal rather than the pathological body, but this is not to say that anatomists of the earlier period had no interest in understanding the structures of disease. Two sixteenth century physicians— Jean François Fernel and Gabriele Falloppio—for instance, produced quite extensive treatises on the causes and structure of tumours, making use of Galenic observations blended with reasoning drawn from other ancients, specifically Aristotle.[21] It was, however, with the work of the Italian anatomist Giovanni Battista Morgagni, who published *De Sedibus, et Causis Morborum per Anatomen Indagatis* (Of the Seats and Causes of Diseases investigated through Anatomy), that the intellectual movement of morbid anatomy began in earnest.[22]

In the massive, five book, *De Sedibus*, an octogenarian Morgagni brought together decades of experience in correlating clinical observations with the findings of the postmortem room. It is difficult to overstate the importance of this work to the development of medical science. Morgagni's intellectual project influenced a generation of physicians and surgeons in Europe and helped build a system of clinical-pathological correlation that became known as the 'birth of the clinic' when it appeared in late eighteenth century Paris.[23] In one way, the work of the Parisian doctors to connect anatomy, disease, and the senses of the clinician using close observation and physical examination, was a continuation of an old Hippocratic imperative to teach students at the bedside of the sick. What was more modern, though, was the spirit of the Enlightenment age of discovery that permeated the teaching of medicine in the Paris hospitals. This is perhaps best exemplified in the words of Antoine Fourcroy, a member of the Revolutionary Assembly and professor of chemistry to the Paris medical school, who urged that students should, 'read little, see much, do much'.[24] Nowhere is this better demonstrated in the teaching and researchers of the Parisians than in the work of René Laennec who developed the stethoscope as an aid the senses to allow himself and his students to 'see' (via the medium of sound) into the diseased chest of the living patient, so opening whole new avenues of knowledge and inquiry as a consequence of experience with the new instrument.[25]

The emphasis on observation and experience at the beside combined with use of instrumentation and postmortem dissection created another great legacy for the nineteenth century practice of medicine, that is, the blurring of the distinctions between physicians and surgeons. The snobbery that elevates the mind over the hand has ancient roots, and the terms themselves show something of this: the word physician derives from the Greek for the study or knowledge of nature *physis*, whereas the root of surgeon, or as it was sometimes rendered into English 'chirurgeon' is found in the term *chirurgia*, a Latinized form of the Greek *kheirourgia* indicating one who worked with his hands—an apprenticed craftsman in other words. Although not absolute—thanks to his outstanding work on the battlefield Paré, for instance, earned royal patronage and considerable fame—distinctions between the scholarly, learned physicians and the trade-oriented barber-surgeons persisted throughout the late Renaissance period. While gentlemen academic surgeons did exist, it is not really until the time of the French Revolution that we can say that the professions were becoming more equal in their status and respective practices.

This relaxing of older boundaries was in part responsible for the increasing popularity of medicine as a career in the eighteenth century. With the rapid urbanization of Europe came a rise in the number of anatomy schools providing the basis of a medical education. One notable school was that established in London by William Hunter, a famed surgeon and obstetrician to the Royal household. It was William's younger brother John though that most concerns us in our story of the prostate, as it was John who wrote several influential pieces about the nature of prostatic disease. It was during his work as a dead-room assistant at the Hunter school that John first developed his flair for the practice of dissection, and John retained his passion as he, along with his brother, became a part of a new generation of gentlemen surgeons.[26] John was also the uncle and instructor of the Scottish physician Matthew Baillie, whose *The Morbid Anatomy of Some of the Most Important Parts of the Human Body* in 1793 may have been dwarfed in the historical record by Morgagni's great treatise but which was, nevertheless, widely circulated and republished in and after his lifetime—helping to spread the ideas of morbid anatomy in the English language.

If we return once again to the prostate, we can see how the work of Morgagni was of such influence to men like John Hunter in both his theory and his practice. Until the time of Morgagni (and aside from the examples described above), the prostate was little mentioned in the medical literature. That said there *was* a great deal of attention paid to a common ailment of men as they aged: problems urinating, also known as strangury.[27] As the life expectancies of city dwellers increased during the eighteenth century, it behoved a doctor to know as much as he could about the likely complaints of his fee-paying patients as they aged. Morgagni's observations in *De Sedibus* of growths from the prostate blocking the neck of the bladder were themselves, he said, based on earlier anatomical finds of some of the great physician anatomists of the late seventeenth century such as Theophilus Bonetus, Marcello Malpighi, and Thomas Bartholin. Morgagni's great contribution, though, was to not just the remarkable feat of placing the diseases of the prostate within a comprehensive account of human morbid anatomy, but also to determine that the 'carnosities' or 'calculi' of the prostate were indeed morbid growths and not variations of normal anatomy. The question was an important one not only for understanding the anatomy underlying disease but also for understanding the structure of the normal prostate particularly as regards to the presence or otherwise of a middle lobe (the middle lobe of prostate, especially when swollen, still today carries the name 'Morgagni's caruncle'), so shifting the anatomy of the gland away from the bifurcated structure supposed by Renaissance anatomists.

Just as Paré had linked disease affecting the prostate with strangury (*chaude-pisses*), two centuries later Morgani, while not suggesting that a diseased prostate was the *only* cause of strangury, urged his fellow clinicians to always be mindful and suspicious of the gland when faced with those symptoms. He emphasized the importance of the prostate in causing diseases in several of the patient cases presented in *De Sedibus*. For instance, in reference to the case of a seventy-five year old man for whom the cause of death was given as 'ascites', or abdominal swelling, Morgagni speculated on the relationship of the prostate to the strangled bladder opening:

> When the anterior portion of the parietes of the bladder was opened longitudinally, I observed in the opposite part a roundish protuberance, the size of a small grape, and covered with the inner coat of the bladder. Its nature I immediately conjectured, and by thrusting my scalpel into it, I divided the projecting body and the prostate gland at the same time, and demonstrated that the former was continuous with the latter. Undoubtedly, had it grown out to a greater degree, it must have become a very considerable impediment to the discharge of urine.[28]

Dwelling on the significance of this case and other observations like it, Morgagni went on to say that he thought such prostatic growths were likely quite common in older men:

> If you attentively examine those examples which I have pointed out ... you will observe that they were all from old men: and in like manner, if you examine my observations in which there was the beginning of a caruncle, you will find that this was found to grow out in the very middle of the internal, and upper circumference of the gland, posteriorly; but whether all these things happened by chance, or otherwise, future observations will show.[29]

In a further revision of the case, Morgagni felt able to settle the issue of whether indeed such things 'happened by chance or otherwise':

Since writing the preceding account I find, that what I have described as an incipient excrescence of the prostate gland, an anatomist of celebrity has considered natural to it, and has described it under the appellation of *uvula*; but I cannot reconcile this with my own observations. In the great number of bodies which I have examined, I have only observed this appearance in the cases which have been alluded to, and in another man, in whose bladder the body in question projected from the posterior part of the orifice of the urethra, in the shape and size of a small cherry; and its structure was evidently an extension of the prostate gland itself.[30]

(The rather withering putdown of Morgagni's fellow 'anatomist of celebrity' was probably the French physician Joseph Lieutaud who had himself published an ambitious four-part series on morbid anatomy in the 1760s.)[31]

In his *A Treatise on the Venereal Disease* first published in 1786, John Hunter followed the lessons of *De Sedibus* in that much of the *Treatise* is devoted to exploring connections between observed symptoms, patient testimony, physical examination, and postmortem findings. The *Treatise* is also where Hunter contextualizes his observations on venereal disease within an expansive survey of the character and consequences of inflammation in all the parts of the penis, testicles, and associated organs and structures. The most important of these consequences was the obstruction of urine leading to bladder distension and even death. As mentioned above, that notion that men, particularly older men, could be susceptible to such a fate was described as far back as the ancient Egyptians, and the use of the catheter easing the discomfort and danger of urine retention dates at least as far back as that time.[32] Similarly, the idea that the male urinary tract was susceptible to 'carnosities', 'caruncles', or 'calculi' dates from at least the time of Galen,[33] as does the insertion of medicines via catheters and other devices to try to reduce the swellings and ease the blockages that they caused.[34] Hunter described his own experience with using bougies to ease strangury:

When a difficulty in making water takes place, a bougie is the instrument which the surgeon will naturally have recourse to, and if he finds the passage clear, which he often will, in such cases he may very probably suspect a stone. If search is made and no stone felt, he should naturally suspect the prostate gland, especially if the ... instrument used meets with a full stop, or passes with some difficulty just at the neck of the bladder. He should examine the gland. This can only be done by introducing the finger into the anus, first oiling it well, placing the forepart of the finger towards the pubes; and if the parts, as far as the end of the finger can reach, are hard, making an eminence backwards into the rectum, so that the finger is obliged to move from side to side, to feel the whole extent of such a swelling, and it also appears to go beyond the reach of the finger, we may be certain the gland is considerably swelled, and is the principal cause of those symptoms.[35]

It seems unlikely that Hunter was the first to examine the prostate in this way—some authors date this practice back to medieval times[36]—but this was probably the first time that a description of this technique of physical examination attained a large audience. (The popularity of Hunter's *Treatise* helped establish a reliable method of rectal examination of the prostate that would become an integral part of the male clinical work-up that endures to the present day.) For the worst cases where bougies and catheters could not help to empty the bladder, Hunter gave his opinions of the commonest types of surgery: cutting through the perineum, over the pubes (both traditional sites for lithotomy, or 'cutting for the stone', an operation to remove bladder and kidney stones that predated Hippocrates that I describe in more detail in Chap. 3), or through the rectum in order to insert a catheter. While all the techniques had considerable drawbacks, Hunter nonetheless professed a cautious optimism that such surgeries could see patients through the crisis of inflammation and be life-saving interventions if the cause of the blockage could resolve itself in the meantime.[37]

Like many of his contemporaries and predecessors then, Hunter understood that swelling of the prostate was common as men aged and he also recognized that distinguishing between the different kinds of strangury was difficult. He tried to get a better understanding of the role of swelling of the prostate by examining the gland in the cadavers of men who had died with the symptoms of urinary blockage and, 'looking upon the mouth of the urethra from the cavity of the bladder'.[38] One practical outcome for such observations, according to Hunter, was that it allowed for more practical designs for catheters and other tools intended to relieve dangerous obstructions to urination. In the section of *A Treatise on the Venereal Disease* titled, 'Of the Treatment of the Swelled Prostate Gland', Hunter records a consultation with a fellow surgeon who claimed to have had some success in reducing the swelling of the prostate by the intro-

duction of a burnt sponge into the urethra.[39] The cutting away of car-
nucles through the use of slicing or pinching instruments inserted into
the urethra were in use from the time of Paré, but it seems that like many
other doctors Hunter was put off by the abundant haemorrhaging that
accompanied these procedures and instead favoured a more conservative
approach.

Hunter was, in general, pessimistic about treatment of the swollen
prostate and noted that although he had tried inserting opiate cylinders
and had recommended sea bathing he really had little confidence in any
of the supposed cures then popular. He was, furthermore, generally dis-
missive of the use of the then commonly used technique of blistering the
perineum to effect a humoral rebalancing (although he did allow for blistering, purging, and bleeding to restore humoral function in other kinds
of inflammation such as that affecting the testicles[40]). It is worth remem-
bering once again that for all of the new ideas and practice in medicine
there remains a great deal of the old, and like many of his contemporaries,
Hunter found ways to interpret the new morbid anatomy within older
Hippocratic and Galenic theories of the humours.

Following John Hunter's death in 1793, his brother-in-law Everard
Home published numerous works based on his studies under Hunter, and
also likely Hunter's own work plagiarized and published under Home's
own name.[41] Home's two volume, *Practical Observations on the Treatment
of the Diseases of the Prostate Gland*, vol. i. 1811, vol. ii. 1818, continued
the tradition of meticulous recording of patient symptoms, signs from
physical examination and, where possible, correlation with postmortem
findings. The volumes were the most extensive treatise ever printed to that
point on the prostate and included a number of what Home claimed to
be his novel discoveries, most prominently the anatomical description of
the gland as having a lobular structure containing a previously 'unknown'
middle lobe, in which inflammation caused the most serious sorts of urinary blockage (as mentioned above, something that had been described
decades before by Morgagni).[42] Nonetheless, Home's attention to the
prostate was without precedent and included an impressive survey of pro-
sected parts—which Home prepared by distending bladders with water—
that would, he hoped, help him better understand the normal anatomy of
the gland.[43]

Like Hunter, Home noted that the likelihood of prostate problems
increased with age, going so far as to say, 'it is a rare occurrence for a man
to arrive at 80 years of age, without suffering more or less under disease

of this part'.[44] Other causes of prostate swelling Home ascribed mainly to inebriety in drink, diet, and relations with women, but none of these were more important as causes as was natural ageing. Home underscored his belief by claiming his observations has a biblical precedent: 'perhaps we may be justified in believing that it is alluded to in the beautiful description of the natural decay of the body, in the Bible, the book of Ecclesiastes, the 12th, chapter, the 6th verse, where it is written, "or the pitcher be broken at the fountain, or the wheel broken at the cistern"'.[45] For sure, the morbid anatomy of the eighteenth century had done a great deal to describe the nature of Paré's *chaude-pisses*, but the line between natural processes, including the seeming inevitably decline in urinary function, and discrete disease process in the prostate remained far from clear. Similarly, while changes to the prostate were clearly common and widely described, what it might mean for the prostate to be 'cancerous' in the early nineteenth century was an open question.

OF INFLAMMATION AND SCIRRHUS: CANCER BEFORE THE MID-NINETEENTH CENTURY

The ancient concept of cancer as articulated by the Hippocratics and Galen was rooted in the notion of inflammation as caused by an accumulation and corruption of the humours. As in other areas of medicine, these ideas of Galen remained little altered until the blossoming of research in anatomy and medicine during the late Renaissance period. Even then, the new ideas of the scientific revolution retained many of the older ideas: Gabriele Falloppio writing at the turn of the sixteenth century, for instance, attempted to more finely distinguish between types of inflammation, but his underlying theory is solidly Galenic. Falloppio's work did further draw some distinctions between 'scirrhus' and 'cancer'—the former, he said, being a hard, but indolent, mass, the latter being a painful, life threatening mass—noting that scirrhus can and does transform into cancer. His overall project was to refine rather than overturn the categories of Galen, which he attempted to do—like other humanists of the period—by bringing an Aristotelian interpretation of matter and form to comment on and modify Galenic principles.[46] The alchemists and iatrochemists of the sixteenth and seventeenth centuries—Sylvius, Paracelsus, and J.B. van Helmont, to name a few—did considerably more to disrupt the centrality of Galenic ideas. Their new ideas of 'vitalism' along with notions of 'distillation' and 'fermentation', for instance, allowed for a fundamentally different framework in which to explore and interpret the body in health and disease in frankly non-Galenic terms.

That the body contained structures and vessels for substances other than blood was an ancient concept. The works of Galen describe lymphatic structures (which he termed *adenón* or 'spongy flesh'), and the idea that the body contained vessels carrying watery or milky fluid (lymph) seems to have been widely accepted by gross anatomists well before the seventeenth century. It was not, however, until the work of two Prussian academics, Georg Ernst Stahl and Friedrich Hoffmann working at the University of Halle at the end of the seventeenth century, that we see a comprehensive study on the role of lymph within the study of morbid anatomy. The pair took a particular interest in the role of lymph in inflammation, and, furthermore, the study of cancer, because cancer was, as Galen had proposed, the worst outcome of inflammation. As with so much else, their ideas then were partly old, partly new, depending both on Galenic conceptions of inflammation as concentrations of humours as well as more modern conceptions of 'corruptions' and 'stagnations' emerging from iatrochemistry.[47] From this later set of ideas, Stahl suggested a kind of cancer 'seed' or ferment that could not only transform scirrhus masses but spread cancer elsewhere in the body: this, he proposed, explained why tumour extractions were seldom effective in saving the life of patient, as the tiny cancer seeds were left behind.[48]

John Hunter took up this lymph theory of cancer in his own researches and proposed a category of 'coagulable lymph', a substance that underpinned all normal function in the body, from growth to repair. Hunter's phraseology emerged from his observations that extracted blood left at rest separated out, with the serum becoming 'squeezed' out from the coagulated part, which (in his terminology) was itself was composed of coagulable lymph and 'red particles'.[49] Such coagulable lymph played a role the body's normal process of blood clotting and healthy inflammation, but it could also, under circumstances of unhealthy or pathological inflammation, extrude from the vessels and produce new, solid structures in the organs and linings of the body.[50] Not all of these new growths were cancers—Hunter recognized warts and polyps, for instance—but he did propose a new categorization of cancer based on his new theory:

> The diseases commonly classed [as cancer] are in appearance very different, and probably are very different in their nature; they should not, therefore, be called by the same name. I would call that cancer which produces the fol-

lowing effects; viz., a circumscribed tumefaction with much hardness, and a drawing in of the skin covering it, as if the cellular membrane underneath was destroyed: then a species of suppuration takes place in the centre, and ulceration of the external surface. This is its most frequent appearance. ... There is another disease which is also called cancer, which I have called fungated ulcer.[51]

Fungated ulcers, Hunter explained, were, like cancer, fatal, but unlike cancer, which ate away at the parts, fungated sores rather threw out spongy fungus matter that could not 'be kept down'.[52] While noting that the fungated ulcer did not seem to 'poison' the surrounding tissue in the way cancer did, Hunter urged their removal, observing as he did so that the fungus would likely return and signal a terminal course: 'It kills without appearing to have done much mischief, whereas cancer does much local mischief, and so do the consequent ones, before death.'[53] By 'consequent ones', Hunter means those cancers that appear secondary to the original tumour due to the 'cancerous poison' spread through the lymphatic fluids, and that may appear at short or long intervals after surgical intervention in the primary tumour.

From Hunter's description of this 'fungus' we likely get the first mention of the 'fungus haematode' in the literature from the British surgeon William Hey, who named it so due to the large admixture of blood and lymph he observed in the excised tumours of his patients.[54] It is the cases of Hey, detailing patients presenting with growths of the breast, neck, and limbs, which Langstaff cites as the basis of his classification of fungus haematode of the prostate. More fundamentally though, Hunter also gave us a vision of cancer—whether 'hard' or 'soft'—that was dependent upon the creation of new growth or 'neoplasm'. Again, though, there is no clear break with ideas of the past—Hunter theorized within a generally Galenic conception of a humoral body, albeit one modified by the new sciences of chemistry and morbid anatomy. The generation of morbid anatomists that came after Hunter would, however, eventually begin the uncoupling of western medicine from its Galenic roots through its continued focus on disease as it manifested in the solid structures of the body.

At the turn of the eighteenth century, Marie-François Xavier Bichat published a remarkable pair of books based on his extensive anatomical researches. In his 1799 *Traité des Membranes* (Treatise on Membranes) and his 1801 *Anatomie Générale* (General Anatomy), Bichat proposed a categorization of the parts constructed on shared textures and appear-

ances. Within this new taxonomy, organs and other structures were understood to be not discrete or unique in and of themselves but rather to be
composed of a combination of entities—membranes, cellular tissue, vol-
untary and involuntary muscle, glands, and so on—several of which might
be found together within the same organ or shared in common between
quite different organ and anatomical sites. The anatomical treatises of
Bichat are all the more remarkable to modern eyes when we consider that
the man refused to use (the by then quite well developed) microscope to
aid his research. Bichat believed the microscope to be more prone to error
than—and so far inferior to—the eye, and one significant consequence
of this element of his work is that what *Bichat* meant by 'cellular tissue'
(*tissu cellulaire*) should not be confused with later definitions of the term.
In naming his structures, Bichat followed the Latin meaning of *cella* as
'compartment': for him, cellular tissue was tissue that could be examined
via insufflation with air due to the presence of internal spaces or compartments (probably more akin to what we would now classify as connective
tissue or fascia).[55] The legacy of Bichat was to demonstrate how to analyse
the body in terms of its basic building blocks, so shifting the investigation
of diseases away from the complicated, gross end-state of disease in whole
organs and towards the more subtle and (potentially, at least) less baffling
'lesions' of the different tissues.[56]

In a wide-ranging research enterprise, Bichat did not study cancer in
any particular depth, but he did explain what he thought his system added
to the question of the nature and causes of tumours. He believed that
tumours all had one thing in common: they were *cellular*, by which he
meant that they were 'overgrowths' of the cellular tissue from which they
emerged. In this view, the differing nature of the tumours—cancers, polyps, fungoids, and so on—depended on the nature of the particular cellular
base and the nature of morbid cause stimulating excess growth (Hunter's
coagulable lymph was suggested as a possible culprit). Followers and critics of Bichat (often, like his famous pupil Laennec, people who were one
and the same individuals) argued about whether tumours manifested in
one kind of way or another depending on the tissue, or whether certain
tissues gave rise to only certain types of tumour. While Bichat's reordering
of pathological inquiry and classification did offer a tantalizing glimpse
of a new clarity in understanding how diseases arose in the body, what
was occurring in the disease process at the fundamental (sub-clinical)
level was still little understood at the middle of the nineteenth century.
While, in Bichat, we have for the first time a theory of disease and tumour

development that rested on the solid structures of the body (an analytical approach that would finally lead to the end almost two thousand years of humoralism), for the diagnosing clinician cancer remained a series of signs and symptoms observed at the bedside (perhaps supported, but not led by the efforts of a microscopist): was the patient's condition a wasting one?; could growths be seen to spread to surrounding (or distant) parts or re-emerge after excision? Only clinical observation could answer these questions during the life of the patient; the analytical power of morbid anatomy seemingly better suited to provide good research questions rather than practical answers, a situation that would only begin to change at the turn of the century when influence of the great 'age of analysis' began to reverberate across medical praxis.

'Analysis', in the sense of separating an entity into its constituent parts, was a style of reasoning that Bichat himself demonstrated in breaking down the complexity of the body into simpler categories of tissues, but the maturation of the approach would go on to coevolve with the rise of the scientific laboratory. Thanks to massive state investment in science as a means to achieve international prestige and industrial know-how, laboratories as a modern reader might imagine them began to emerge in the Germanic world during the early nineteenth century. References to 'laboratories'—a term that, depending on the source, seems to be derived either from a Latin neologism to denote a 'place to work', or else as a form of 'elaboratory' from the Latin form of 'elaborate', used as a verb— are found in English, French, Spanish, Portuguese, and Italian texts from the sixteenth century onwards, usually in reference to the work spaces of alchemists. By the seventeenth and particularly eighteenth centuries, private laboratories—perhaps a converted room in the home—had become fashionable playthings of interested amateurs. During the same period, laboratories also began to appear in museums of botany and anatomy as those institutions themselves began to become important sites of teaching and research (as seen with the museum and laboratory of the Hunters of London, for instance), but the idea of the laboratory as an *instrumental* part of the industrial and academic enterprise has considerably more recent roots.[57]

In a departure from the earlier types of 'laboratory', in the nineteenth century Germanic sense, the 'new' version of the purpose-built laboratory emerged as a rather ambitious, highly cooperative kind of enterprise explicitly designed to disseminate its styles of practice by training new professionals and opening up new lines of enquiry to nurture careers at the nascence of academic science and medicine. The most famous early model of this type was Justus von Liebig's laboratory for animal chemistry established at Giessen in the 1820s, but other contemporary researchers, notably Johannes Müller in physiology and Ernst von Baer in embryology, were similarly active in bringing new approaches across a swathe of questions in the life sciences. The protégés of this first generation of researchers—investigators like Karl Ludwig, Theodore Schwann, and Rudolf Virchow—would come to establish German laboratory science as an international force by the mid-nineteenth century. The close connections between the universities and industry encouraged by the state further provided new tools and technologies for investigation in the form of chemicals and machinery, but also in what would become the iconic symbol of the German laboratory: the microscope.[58]

In the seventeenth and eighteenth centuries, Antoni van Leeuwenhoek, Robert Hooke, and Marcello Malpighi (whose observations of microscopic capillaries finally provided the evidence for William Harvey's theory of the circulation of the blood, albeit more than a decade after Harvey's death) all conducted famous studies of the natural world using the microscope but the widespread uptake of the instrument was slow. Like Bichat, many investigators preferred the evidence of their eyes to the blurred and distorted images they saw through microscopes. During the early nineteenth century Joseph Jackson Lister (the father of the antisepsis pioneer, Joseph Lister) designed a lens that dramatically reduced aberration effects, so producing higher quality resolution of images and a means to challenge the widespread scepticism surrounding the microscope.[59] Using his new lenses, Lister and fellow English physician, Thomas Hodgkin, developed a classification of tumours of the glands that endured (eventually becoming eponymously named for Hodgkin) but the histological basis of their claims was controversial and was soon overturned. Beyond these individual efforts, however, the command-economy industrial boon that had simulated the growth of laboratory science on the continent was now also creating new tools and techniques in manufacturing and production, specifically in the case of microscopy, with the industrialization of the German lens makers—Zeiss and Leitz—that transformed the attitudes to the use of microscope through high-quality mass production by the mid-to-late nineteenth century. It was these two things together then, the advance in technology combined with the new institution of the laboratory (with its associated freedoms and opportunities for would-be career scientists), which established a new era of innovation in medical research.

Using the new microscopes, Virchow, along with Schwann and others, developed an understanding of the 'cellular' to replace the 'compartments' of Bichat. They literally and figuratively viewed the building blocks of life through an extraordinary new lens. As the historian William Bynum puts it: 'Virchow's *Cellular Pathology* (1858) did for the cell what Morgagni's *Seats and Causes of Disease* (1761) had done for the organ, or Bichat's *Treatise on the Membranes* (1800) had for the tissues', that is he (Virchow), 'established a new, essential unit for thinking about function and disease'.[60] As usual, though, transformative ideas did not emerge neatly so much as they resulted from a messy contest and combination of the novel and the established. Virchow's departure from several of his fellow cell theorists was in his insistence on the materiality of function in health and disease, for instance. His close colleague, Schwann, on the other hand, followed a kind of humoralism (similar to Hunter's coagulable lymph), advocating that cancer formed out of a 'blastema' or structureless substance found between cells irritated to pathological changes by substances appearing in the blood.[61] Virchow's now famous phrase, *omnis cellula e cellula*, or 'each cell from a cell', compellingly links the microscope to a material, cellular, vision of health and disease. Given the high profile contemporary neo-humoralists like Schwann, however, is was not necessary obvious or inevitable that Virchow's vision of pathology as something occurring within cells, not humors, would come to dominate the new academic medicine.

As Virchow's theories were woven into the great tapestry of ideas of the long nineteenth century, so the central challenge of the age remained: how to draw together clinical observation, morbid anatomy, clinical-pathological correlation, and now histology, to determine ever-finer distinctions between diseases (so creating a robust nosology) while also seeking out clues about effective therapeutic intervention through the uncovering of underlying mechanisms. This was an extremely difficult task, of course, and one that tested (and confounded) the abilities of generations of physicians, surgeons, and clinical researchers. Within this conundrum, the classification of tumours was perhaps particularly gruelling. As I described earlier in this chapter, the many textures and types of tumourous growths were well described by the Ancients, but by the nineteenth century medical commentators were looking to make sense of and rationalize these descriptive categories, not to further add to the prolixity. Trying to determine what was a discrete entity versus a mere variation, or, crucially, what was a cancer and what was a tumour of another kind, was frustratingly elusive. As the English surgeon John Abernathy put it in 1804 when arguing about whether sarcomas were, like carcinomas, properly consider as 'cancers', the task itself was undermined by the lack of clarity of what that old category even meant anymore: 'The term cancer is objectionable,' Abernathy concluded, 'because it conveys an erroneous idea of its nature; for this disease, though perhaps equally destructive, will be shown to be unlike cancer in its nature and progress'.[62] So did Virchow fair any better? Using very different tools and concepts decades later, he did attempt an improved characterization of cancer and the tumours. In his work on the prostate, for instance, Virchow to suggest that 'adenoma' be distinguished from 'myoma' (arising from the gland or from the surrounding muscular tissue, respectively), but he, too, remained in doubt as to what did and did not constitute a 'true' cancer.[63] Ultimately, as the historian Carsten Timmermann puts it, 'microscopy and cellular pathology added new interpretative devices to the toolkits of physicians, surgeons, and pathologists, but they continued to [also] use the old ones'.[64]

The first two articles to attempt a comprehensive overview of prostate cancer—both published in the 1860s—are a testament to the pluralism of later nineteenth century clinical investigations. 'Die Heterologen (Bösartigen) Neubildungen der Vorsteherdrüse' (The Heterologous [Malignant] Tumours of the Prostate Gland) authored by the Polish clinician, Oskar Wyss in 1866, for instance, was part historical literature review, part collection of case histories of patients he had observed during his work at the Breslau Clinic, and part the results of his own and others histological studies of tumours. The article was published in *Archiv für Pathologische Anatomie und Physiologie und für Klinische Medicin* (Archives of Pathological Anatomy and Physiology and of Clinical Medicine), a journal founded by Virchow (and later renamed in his honour) and ultimately rested on a very modern (very Virchowian) delineation between tumours composed of glandlike structures resembling the prostate gland itself, and tumours composed of epithelial like cords resembling the connective tissue independent of the gland.[65] Similarly, when *Essai sur le Cancer de la Prostate* (An Essay on the Cancer of the Prostate) appeared a few years later in 1869 written by the Parisian internist Jacques Jolly, the review of the literature was still more lengthy and comprehensive including many more case studies and more morbid anatomy alongside observrations drawn from the new cellular theory.

Jolly, in particular, vigorously engaged with numerous debates of his time, including the possible infectious nature of cancer (a resonant fear in

the increasingly crowded cities), as well as speculations over the true incidence of prostate cancer. On this latter point, he noted that most authors
from Morgagni onwards agreed that the commonest urinary tract problems were caused by prostates engorged by scirrhus, but that scirrhus in
this 'classical', or ancient, sense was not (again) easily or readily translated
into a clinical categorization of cancer. As Jolly pointed out, induration
(hardening) of the prostate could be called 'scirrhus', but not all scirrhi
were cancer.[66] On Langstaff's observation of fungus hematodes (soft cancer) of the prostate, Jolly again expressed uncertainty about their pathological nature.[67] A further area of confusion voiced by Jolly was how to
determine where the cancer originated: how to tell if a tumour had really
emerged from the prostate, rather than, say, spread there from the bladder or the testicle or some other structure? So while Jolly did try to bring
histological insight into resolving the nature and origin of prostate cancer,
he was deliberate and conservative in his conclusions. For instance, while
he was sympathetic to Wyss' classifications (following from Virchow), he
urged caution based on his own observations that the gland of a patient
deceased with prostate cancer was usually in a state of 'complete disorga-
nization', providing the observer with the hugely complicated task of trying to discern the regular tissue, diseased tissue, and the point of 'morbid
production'.[68]

Jolly, Wyss, Abernathy, Langstaff, and Adams, along with other clinicians of their era interested in urinary diseases, were then profession-
ally preoccupied with distinguishing between 'ordinary' scirrhus and the
'true' cancer, whether degenerated scirrhus of the prostate or some kind
of fungating tumour. While many commentators were content to dismiss
scirrhous and fungating prostate as vanishingly rare, Jolly urged caution
noting that confusion in classification of cancer and non-cancer made generalizations from the existing literature difficult, as did the fact that few
clinicians even *knew to suspect* a case of cancer (a state of affairs not at all
helped by the prevalent view of cancer as being a very rare disease).[69] Jolly
did highlight one important area of agreement seemingly shared by most
clinicians, that in addition to the chronic retention of urine and occasional
urtheral haemorrhage seen in ordinary scirrhus of the prostate, *true* cancer
involved pain in urination, cachexia (physical wasting) and pains in the
perinium and back, so could be diagnosed clinically with reasonable accuracy. [70] He acknowledged that divisions of opinion remained on almost
every other basic question—even over which patient groups were most
affected. Was it the case as some claimed that prostate cancer was a disease

of children, and not as might be assumed of older men? Could it be found across age groups? Jolly's opinions on these issues was firmly for the view that prostate cancer was found almost exclusively in older men, but the fact he felt he needed to address the issue at all is illustrative of how, in the case of cancer and diseases of the prostate, questions were many and definitive answers few.

To conclude then, we can see that the early-to-mid nineteenth century was a time in which the newly drawn structures of disease were still largely interpreted within a much older concept of humoralism in which disorders of the fluids provoked what Paré might have described as the 'many mis-chieves' of disease. With little to say about the causes, or aetiology of dis-ease, and still less to offer as regards to treatments, the morbid anatomists instead concentrated their efforts on exposing as precisely as they could the inward devastations of the outward symptoms. Like the Hippocratics, these investigators recognized the severe limitations of their ability to intervene in the natural processes of the world, but like the Hippocratics they had an unwavering confidence that their researchers would one day lead to transformative knowledge. As the most famous of the Hippocratic aphorisms put it, 'Life is short, science is long; opportunity is elusive, experiment is dangerous, judgment is difficult.'[71] Such are the perennial challenges of research. As the nineteenth century wore on and diseases of the prostate became better understood, another generation of doctors began yet a new kind of experimental endeavour, this time in the realm of surgery.

NOTES

1. Quoted in Murphy and Desnos, *The History of Urology*, 65.
2. Lawrence, Cases of Fungus Hæmatodes, 279. A bougie is a thin, sometimes flexible, cylinder used for the purposes of dilation.
3. Ibid., 281–2.
4. Adams, The Case of Scirrhus of the Prostate Gland, 393.
5. Ibid.
6. Ibid., 394.
7. Hippocrates et al., *Hippocratic Writings*, 248.
8. Ibid., 247.
9. Ibid., 49.
10. Rather, *The Genesis of Cancer*, 9–10.
11. Hippocrates et al., *Hippocratic Writings*, 140.
12. Ibid., 170.

13. Ibid., 230.
14. Josef Marx and Karenberg, History of the Term Prostate, 209.
15. Nutton, *Ancient Medicine*, 222–5.
16. Rather, *The Genesis of Cancer*, 13.
17. Galen, *On Examinations by which the Best Physicians are Recognized*, quoted in Von Staden, Anatomy as Rhetoric: Galen on Dissection and Persuasion, 56.
18. Rutten, Early Modern Medicine, 65.
19. Josef Marx and Karenberg, History of the Term Prostate, 210.
20. du Laurens, *Historia Anatomica Humani Corporis*.
21. Rather, *The Genesis of Cancer*, 15–19.
22. Porter, *The Greatest Benefit to Mankind*, 263.
23. Bynum, *Science and the Practice of Medicine in the Nineteenth Century*, 28.
24. This period still remains best described by Ackerknecht's classic, *Medicine at the Paris Hospital, 1794–1848*.
25. Reiser, *Technological Medicine*, 1–13.
26. For a discussion of the rise of the gentleman surgeon, see Porter, William Hunter: A Surgeon and a Gentleman.
27. Murphy and Desnos, *The History of Urology*, 82–6.
28. Morgagni and Cooke, *The Seats and Causes of Diseases: Investigated by Anatomy*, 2:304.
29. Ibid., 2:305.
30. Ibid.
31. Thompson, *The Diseases of the Prostate*, 20.
32. Moog, Karenberg, and Moll, The Catheter and its use from Hippocrates to Galen, 1196.
33. Murphy and Desnos, *The History of Urology*, 63.
34. Moog, Karenberg, and Moll, The Catheter and its use from Hippocrates to Galen, 1197.
35. Hunter, *A Treatise on the Venereal Disease*, 171–2.
36. Shelley, The Enlarged Prostate, 470.
37. Hunter, *A Treatise on the Venereal Disease*, 184–91.
38. Ibid., 170.
39. Ibid., 175.
40. Ibid., 91.
41. Payne, *With Words and Knives*, 146, n. 56.
42. Home, *Practical Observations on the Treatment of the Diseases of the Prostate Gland*.
43. Ibid., i:8–9.
44. Ibid., i:18.
45. Ibid.
46. Rather, *The Genesis of Cancer*, 15–16.

47. Ibid., 34.
48. Ibid., 35.
49. Hunter, *Lectures on the Principles of Surgery*, 30.
50. Ibid., 151.
51. Ibid., 367.
52. Ibid., 387.
53. Ibid., 388.
54. Hey, *Practical Observations in Surgery*, 152.
55. Rather, *The Genesis of Cancer*, 54.
56. Bynum, *Science and the Practice of Medicine in the Nineteenth Century*, 32.
57. For an overview of the history of the term and concept of the laboratory see Gooday, Placing or Replacing the Laboratory in the History of Science?.
58. On this period in German laboratory science see Otis, *Müller's Lab*.
59. Reiser, *Medicine and the Reign of Technology*, 69–76.
60. Bynum, *Science and the Practice of Medicine in the Nineteenth Century*, 100.
61. Porter, *The Greatest Benefit to Mankind*, 330–1.
62. Abernethy, *The Surgical Works of John Abernethy*, 56.
63. Murphy and Desnos, *The History of Urology*, 384.
64. Timmermann, *A History of Lung Cancer*, 24.
65. Rather, *The Genesis of Cancer*, 145.
66. Jolly, *Essai sur le Cancer de la Prostate*, 5–6.
67. Ibid., 6.
68. Ibid., 38.
69. Ibid., 12.
70. Ibid., 43–44.
71. Hippocrates et al., *Hippocratic Writings*, 206.

CHAPTER 3

Surgery and Specialization

I operated on a patient of Mr Forbes, surgeon at Camberwell, and removed an immense number of prostatic calculi. These calculi had produced not only painful feelings in the perineum, but a degree of irritation, which kept the patient in continued mental excitement, bordering on insanity. I introduced a staff into the bladder through the urethra, and opened the perineum as far as the prostate, cutting into the urethra, as in the operation of lithotomy; I then made an incision into the left lateral lobe, and extracted many calculi from a bag formed in it. … The sufferings of the patient induced me, about six months after the first operation, to perform a second … : I extracted about half as many calculi as in the first operation. The patient soon recovered from the effects of this second operation, and the wound closed entirely; but, after a short time, his sufferings became as dreadful as before, and, believing that he could not procure any relief, he destroyed himself six months after the second operation. Astley Cooper, Of Calculi in the Prostate Gland (1831)[1]

When it is palpable, well above the prostate, even though the upper portion of the seminal vesicles seem uninvolved, the chances of radical cure are not good. Where a line of induration extends upward and outward, and particularly if any enlarged glands are felt well out along the pelvic wall, the prognosis is also very bad. In such cases a radical cure can not be expected. Metastases to the bones of the pelvis and spine are important to recognize. Hugh Young, Radical Cure of Carcinoma of the Prostate (1904)[2]

The white heat of the Paris clinics helped in many ways to create a new 'modern' style of medicine: one that blended medical and surgical ideas, was research and teaching intensive, and which was, in other words, an

© The Editor(s) (if applicable) and The Author(s) 2016   43
H. Valier, *A History of Prostate Cancer*,
DOI 10.1057/978-1-137-56595-2_3

early version of the phenomenon that we would now describe as 'academic medicine'. An important feature of this new trend was the extent to which 'specialist' knowledge and practices could be rapidly developed. Take the career of Pierre-Joseph Desault, for instance, a man who began his professional life relatively inauspiciously as a barber-surgeon but who would go on to found famous centres for academic surgery, first at the Charité then at the Hôtel Dieu in late eighteenth century Paris.[3] At the College of Surgery, Desault oversaw the building of a grand surgical amphitheatre that allowed students to observe and be instructed on live surgeries.[4] A similar innovation on the wards, where Desault insisted that surgical students be placed in charge of wards—caring for patients and maintaining records—was an enormously important step, one that opened a new world of practical opportunities for would-be practitioners just as their elevation in status itself symbolized the ascendant status of academic surgery as a whole.

This new spirit of enthusiasm for research and instruction did not, however, necessarily extend to attitudes towards therapeutic outcomes. When John Hunter's protégé, Astley Cooper, wrote up his tragic case of prostatic calculi in 1831 he did so at a time of modest expectation for the likely success of urologic intervention. Driven by a patient's pain or chronic obstruction of urine, a surgeon like Cooper might operate to provide what relief he could while acknowledging that the chances of permanent remediation were low and the risks high. The uptake of general anaesthetics around that time *did* go on to help usher in a more hopeful mood in surgery more generally, but it was a mood tempered by the persistent dual-threat of surgical accident and post-surgical infection. Nevertheless, by the end of the century several prominent surgeons were encouraging more extensive, more ambitious kinds of intervention, and no one individual better exemplified this new age of optimism than the brilliant and ambitious figure of Hugh Young. As a faculty member at the new bastion of academic medicine in the US, Johns Hopkins School of Medicine (established in 1893 based on a model inspired by German academic medical clinics), Young was well placed to experiment with new approaches. One of these was the practice of so called 'radical' surgery something that he, along with some of his colleagues—notably William Halstead, the creator of the radical mastectomy as a treatment for breast cancer—hoped would go well beyond providing limited and temporary relief and instead provide a truly substantive and corrective intervention in the disordered physiological systems of the patient.

For much of the nineteenth century, the pursuit of academic medicine by the US medical community lagged well behind the Europeans in scope and content—it was quite common, in fact, for an ambitious American student to complete his or her education by touring the great universities of France, Britain, and Germany. By the last decades of the century, however, this tide was on the turn. Still, when a study commissioned by the philanthropic Carnegie Foundation was published in 1910, the report— *Medical Education in the United States and Canada*—authored by the educational reformer Abraham Flexner, bemoaned the weak state of North American medical education in comparison to the academic institutions of Europe.[5] While the report did cause something of a stir, the problems that Flexner laid out were generally well known and long recognized within the American medical community itself. Indeed, the 'gold standard' of US academic medicine that the report exhorted as a model to be emulated across the nation was Johns Hopkins, an institution founded precisely to address the kinds lacunae 'exposed' by Flexner.

When the American Medical Association (AMA) formed in 1847 it did so in part as a response to the problems (especially as perceived of by younger, less established, and less well-connected community of physicians) of a crowded medical marketplace.[6] In the first few decades of its existence the AMA had only limited impact, and the growth of 'alternative' medical sects like homeopathy, hydropathy, chiropractic, and Thomsonianism flourished. One way that the AMA responded to these pressures was to emphasize how the new laboratory-based curricula of the elite Europe school should form the basis of a new way of educating the American doctor (and so a new way of defining what the profession was and was not, and who was inside and who was outside of its boundaries). It was this emphasis, in turn, that helped nurture a new tradition of specialty practice in the US (and elsewhere) then becoming a popular means for the young and ambitious doctor who might lack access to the comfortable, gentlemanly circles occupied by the established medical elites. The traditional basis of elite knowledge was not easily swept aside, however, and pressing questions concerning the scope and benefits of specialist *versus* generalist praxis and of the 'art' *versus* the 'science' of medicine were vigorously debated. That this should have been the case is hardly surprising: these were distinctions that, in a profound sense, went to the very heart what it meant to *be* a physician in an age of increasingly sophisticated instrumentation and technology. The notion that medicine could and should be taught and 'known' through universal principles synthesized from the analyses of the laboratory sciences, sat in uncomfortable, sometimes dyspeptic, relation to an older ideal of medical knowledge as grounded in observation at the bedside. For its critics, academic medicine appeared to undermine the importance of experience and acumen developed through observation of the specific and idiosyncratic ways that diseases manifested in individual patients. The supporters of academic medicine, on the other hand, for their part saw such traditional features of practice as somewhat necessary but by no means sufficient to progress the state of knowledge.[7]

In addition to these broader debates, each community of specialist practitioners had its own, more specific concerns. In the case of urology, Young's rosy outlook was not necessarily shared by many of his peers—especially those outside the walls of his home institution. As I discuss below, some voiced their concerns that the procedures championed by Young were simply well beyond the technical abilities of most 'ordinary' surgeons. His critics warned that the consequences of too many inept hands taking on his techniques could be dire. Botched operations could easily endanger the health of patients, of course, but they could also threaten the reputation of the (nascent) specialty if and when such failures became widely known. While Young himself never really accepted the validity of *this* particular criticism, he did very much understand what was at stake for the community. He knew, for instance, that in cases of serious and advanced disease—like prostate cancer—the outlook for patients was bleak, but he thought that great improvements in prognosis *could* come if only something could be done about the often lengthy delays between a man first experiencing symptoms and his finding his way to an able surgeon. This delay, he felt, could be best addressed by raising the general awareness of urologic diseases as both common and treatable, but *only* if family practitioners and general surgeons did not hesitate to make appropriate and expeditious referrals. Here again then, the status and reputation of urology was critical: referrals were unlikely to come if confidence in the specialty was low.

For all that was new about the emerging communities of specialists, for urologists at least, their ideas and practices had some very old roots. As discussed in the previous chapter, the problems of the urinary tract in males had been widely debated by learned healers for thousands of years before Young's time. Similarly, the principal instruments of the modern urologist—such as the bougies and catheters—were themselves only slight variations on ancient tools. What I have not yet addressed is the equally ancient technique of therapeutic *cutting* known as 'lithotomy' to extract bladder (and probably kidney) stones. As described in the quotation from Young at the opening of this chapter, this, too, would become absorbed by a community of academic urological surgeons keen to investigate and exploit the procedure for conditions well beyond that for which it was originally intended.

In addition to the forcing of catheters and other rigid structures past urinary blockages, the practice of lithotomy is one of the first recorded operations appearing in documents from the ancient Egyptians to the early Hindus. By the time the Hippocratic *Corpus* began to be assembled in the fifth century BCE, the procedure was common enough for the author of the Hippocratic *Oath* to include a warning against the practice, urging his fellow physicians to 'leave such procedures to the practitioners of that craft'.[8] It is understandable that physicians, particularly physicians as relatively conservative as the Hippocratics, would avoid these operations (which were, in any case, the type of *chīrurgia* or handwork eschewed by 'learned' physicians) due to the risks involved to the patient. A common method of extraction of the stones was via an incision in the perineum, which, apart from the obvious pain and problems with bleeding and infection, could easily puncture or otherwise damage the bladder, resulting in a rapidly fatal outcome. Still, as the writer of *The Oath* noted, some practitioners did specialize in the craft, and as we have an abundance of historical evidence showing the development of quite ingenious tools and procedures across different parts of the world for the next two thousand years or so, the operation, as desperate as it might seem to modern eyes, was in enough demand to be commonly practised in different and competing variations.[9] The acute suffering of patients, such as that described by Cooper, give us a powerful insight on how and why this operation become one of the most commonly practised in human history.

The intellectual and practical developments of surgery discussed in Chap. 2 ushered in a new era for lithotomy (as well as for the practice of crushing stones in place, known as lithotrity) as the ability to practise dissection enabled new techniques to be developed, including that of 'suprapubic' (that is, just above the pubic bone on the lower abdomen) incisions and extraction. Although the suprapubic method had been used sporadically before, most practitioners heeded the Hippocratic prohibition against such a dangerous operation on the bladder.[10] As the study of academic anatomy and surgery grew up in the eighteenth century, so these interventions for stone historically shunned by the learned elites began to be of particular interest to a new generation of surgeons interested in the male urinary and reproductive tract. One early manifestation of this was the early-to-mid nineteenth century fascination with designing new instruments and procedures to go along with them, which frequently led to bitter disputes over priority in discovery.[11] Catheters, dilators, sounds, bougies, and urethrotomes (instruments for cutting the urethra from the inside) were all designed and redesigned as understandings of the structure of the urethra and the nature of its pathologies became more advanced and as new materials—like flexible rubber, for instance—began to be used. Early attempts to visualize the bladder and urethra by means of an illuminated endoscope were also attempted, during the 1820s and 1830s, although the realization of this technique would be several more decades in the making.[12]

As the availability of human remains for clinical research increased during the nineteenth century, so the ability to experiment and demonstrate these new instruments and procedures on the corpse became more widespread. The practice of human anatomy was not the only historical driver for this new confidence in surgical intervention, however: the field of (cross-species) comparative anatomy that blossomed around the same period was at least in part also responsible. To understand why, it's helpful to again look back at the work of John Hunter and his articulations between not only morbid anatomy and surgery but also the science of comparative anatomy. The concept of comparative anatomy had, like lithotomy, ancient roots. Several works of Aristotle, for instance, dealt with classifying animals according to the structures and inferred functions of their skeletons, tissues, and organs. During the anatomical renaissance, Hieronymus Fabricius, professor of anatomy and surgery at the University of Padua, revived this Aristotelian practice while also continuing the more recent traditions of the highly detailed, structural anatomy of Vesalius. Fabricius' teaching in turn influenced the great English physician William Harvey, who would go on to publish the first comprehensive account of an active, heart-based, whole body circulatory system in his *De Motu Cordis* (On the Motion of the Heart) in 1628. Harvey carried out extensive anatomical researchers on all manner of animals but it was by observing the (compared to humans) simpler circulatory systems of fish and eels that he was about better to understand what he was seeing in the body of the human.

Harvey's findings and theories were the stuff of great medical controversy throughout the seventeenth and eighteenth centuries, but in this process comparative anatomy became a highly visible entry into important medical debates. Take, for example, arguments concerning the nature of health and disease. Was it, as some claimed, a kind of hydraulic model of Galenic humouralism in which fluids and solids interacted in the body and could be understood more or less in mechanical and chemical terms? Or was the body animated by some vital force, a soul that nurtured and nourished the body? One major purpose of Hunter's extensive anatomical collections was to settle the debate (he was on the vitalist side), in a controversy that swept up some of the most prominent figures in Enlightenment-era natural philosophy, George Ernst Stahl, Herman Boerhaave, and René Descartes among them. Hunter thus took up a study and a practice of anatomically based surgery and morbid and comparative anatomy at a time when such work was imbued with immense philosophical and practical importance. His remarkable findings concerning the prostate have in retrospect been the proposed starting point for a new science of disease, that of hormone-dependent pathologies. As with most origin stories, however, the progression is far from clear and Hunter's work on the reproductive organs, as groundbreaking as it was, appeared to mostly sink from sight until the end of the nineteenth century.

HUNTER'S COMPARATIVE ANATOMY AND OBSERVATIONS ON CASTRATION

It was by invoking comparative anatomy that John Hunter dryly noted the potential problems inherent in the 'dual-purpose' penis:

It may be observed, that the urethra in man is employed for two purposes. On this occasion I may be allowed to make the following general remark, that Nature has not been able to apply any one part to two uses with advantage, as might be illustrated in many instances in different animals. The animals, whose legs are contrived both for swimming and walking, are not good at either, as seals, otters, ducks, and geese. The animals also, whose legs are intended both for walking and flying, are but badly formed for either, as the bat. ... [A]nd whenever parts, intended for such double functions, are diseased, both are performed imperfectly. This is immediately applicable to the urethra, for it is intended as a canal or passage, both for the urine, and the semen.[13]

More fundamentally, Hunter's studies of comparative anatomy pointed to the deeper insight that all the parts of male anatomy concerned with reproduction were all in some way dependent upon the testicles:

> From these observations it is reasonable to infer, that the use of the vesic-ulae in the animal oeconomy must, in common with many other parts, be dependent upon the testicles. For the penis, urethra, and all the parts connected with them, are so far subservient to the testicles, that I am persuaded few of them would have existed if there had been no testicles in the original construction of the body; and these would have been so formed as merely to assist in the expulsion of the urine. To illustrate this opinion, let us observe what is the difference between these parts in the perfect male, and in a male that has been deprived of the testicles when very young, at an age in which they have had no such influence upon the animal economy as to affect the growth of the other parts. In the perfect male the penis is large ... [o]n the contrary, in the castrated animal the penis is small...[14]

The use of castration was not, historically speaking, limited to livestock, of course. There is a long history of castration from courtly life in the ancient dynasties of China, through to its use as means to control African slaves brought the Americans in the modern period, or, in the later period, to treat various afflictions (e.g., abscesses) of the testicles. When the testicles were severed before puberty as they were, for instance, in the operatic world as a means to produce a male singer capable of singing contralto or soprano (the 'castrati'), the physical results were well known and commented upon. Hunter, in other words, had ample common knowledge at his disposal when thinking on the obvious effects of castration, most particularly early castration, on the development and appearance of the penis. Hunter's detailed anatomical researches, however, allowed him to speculate on the substance of the connection:

> The prostate gland, Cowper's glands, and the glands along the urethra ... are in the perfect male large and pulpy, secreting a considerable quantity of slimy mucus, which is salt to the taste; it is most probably there for the purpose of lubricating those parts, and is only thrown out when in full vigour for copulation: while in the castrated animal these are small, flabby, tough and ligamentous, and have little secretion. From this account there appears to be an essential difference between the parts connected with generation of the perfect male, and those which remain in one that has been castrated;

more especially if that operation had been performed while the animal was young.[15]

Hunter proposed a role for the prostate (along with the other glands lining the urethra) that supported his larger argument that, contrary to the views supposed by earlier thinkers, neither the seminal vesicles nor epididymis nor any other structure acted as a reservoir for semen, but rather,

that it is secreted at the time in consequence of certain affections of the mind stimulating the testicles to this action: for we find, that if lascivious ideas are excited in the mind, and the paroxysm is afterwards prevented from coming on, the testicles become painful and swelled ... pain and swelling is removed immediately upon the paroxysm being brought on...[16]

Hunter continued his essay with a detailed anatomical explanation of how an erection occurs (based on his findings from experiments where he injected fluids and applied ligatures to the penises of dogs), followed by speculation on the role of the various parts in the process of ejaculation. These same observations also confirmed to Hunter under controlled conditions that the removal of the testes did indeed diminish the prostate both in appearance and in function.

Astley Cooper perfected many of his mentor's operations and made contributions of his own, including (of most relevance to this topic) the description of 'malignant' growths of the testicles, fungus hematodes, and scirrhus, and the development of a procedure for surgical castration. Why then didn't Hunter himself propose a series of experiments to castrate men in order to shrink their troublesome prostates? To begin answer this question it is worthwhile to understand what being a professional surgeon meant to him. In his introductory lecture 'Principles of Surgery' Hunter instructed his students to be aware of how their chosen profession was changing:

Too much attention cannot be paid to facts; yet too many facts crowd the memory without advantage, any further than that they lead us to establish principles. By an acquaintance with principles we learn the *causes* of diseases. Without this knowledge a man cannot be a surgeon. Surgeons have been too much satisfied with considering the effects only; but in studying diseases we ought not only to understand the effect, as inflammation, suppuration, &c., but also the cause of that effect; for often without this knowledge our practice must be very confined, and often applied too late, as in many cases it will be necessary to prevent the effect.[17] (original emphasis)

In keeping with the Hippocratic tradition, Hunter urged caution in these new days of surgical confidence, specifically in the ability of a surgeon to recognize the limitations of practice:

> If the disease is already formed, we ought to know the modes of action in the *body* and in the *parts*, in their endeavour to relieve themselves; the powers they have of restoring themselves, and the means of assisting these powers. Or, if these prove insufficient, we judge, by all the attending circumstances, how far excision may be necessary, and what condition of the constitution is most favourable to the operation. To determine on this last point is exceedingly difficult, and in some instances exceeds our present knowledge. I lately saw a patient die in a few hours, of no other operation than the excision of a small tumour from the arm; another, by the removal of one from the abdomen; a third, by castration ... Now all these patients seemed previously in good health, and perhaps the cause of death, and particular circumstances of the constitution which render operations this hazardous, will never be understood.[18] (original emphasis)

Such sentiments—while clearly not dissuading Cooper and others from castration surgery in cases of diseased testicles—may go some considerable way to explain why Hunter and others did not extrapolate from the animal findings and attempt to treat the enlarged prostate via this method.

In 1859 a Parisian physician, Louis Auguste Mercier, did publish one account of an apparently successful surgical castration in the treatment of urinary blockage due to an enlarged prostate, and cited Hunter as rationale.[19] Mercier's account, however, was among only a handful of recorded cases in the decades after Hunter's death. The use of castration in surgical practice would not gain acceptance until the very end of the nineteenth century when the combination of anaesthesia and asepsis encouraged more aggressive surgical approaches. During the mid-1890s, a surgeon at the London hospital, C. Mansell Moullin, became briefly entangled with the University of Pennsylvania professor of clinical surgery, J. William White, over who had been first to resurrect Mercier's operation. The more substantial publication was that of White who in 1895 wrote up a lengthy series of cases in which castration (performed by himself or by his correspondents) had been used to apparently good effect in the treatment of prostatic enlargement and urinary blockage. White was keen to point out that the death rate expected from the operation—between seven and fifteen per cent depending on how the statistics were collected—should not overly concern other surgeons thinking of taking up the technique. He did, though, make it clear that the surgeon might have some persuading to do as patients felt, 'real, entirely natural, and very strong repugnance to the operation',[20] especially if they were younger.

White's paper provoked quite an upset within the tight-knit community of American academic surgeons of which he was a part. In 1896 the Surgeon to the Massachusetts General Hospital, Arthur Tracy Cabot, used his address to the American Surgical Association meeting to roundly condemn White's claims.[21] Cabot disputed White's mortality figures (and those of Moullin which he derided as close to twenty-eight per cent). Cabot's most compelling criticism of castration, however, was that it did *not* reliably shrink the prostate and even when it did it was merely as good as *his* favoured procedure, prostatectomy (the origins of which I say more about below). While it might not have been particularly obvious at the time, this spat was one of several that would soon prove central to the formation of an identity for urology as a surgical specialty.

UROLOGY AND PROSTATE CANCER IN THE MID-TO-LATE NINETEENTH CENTURY

When Theodor Billroth was appointed as professor of surgery to the University of Vienna in 1867 he was already a famous man. His 1863 textbook, *Die Allgemeine Chirurgische Pathologie und Therapie* (General Surgical Pathology and Therapy), was by then published in multiple editions and was on the way to being translated into ten different languages.[22] A major attraction of the text was Billroth's unabashed celebration of the new status of surgery as (at least) the professional equal of medicine. He was keen to point out that the true surgeon must master the arts of medicine if he is to master his own craft: 'In short, the surgeon can only judge safely and correctly of the state of his patient when he is at the same time a physician.'[23] Indications for surgery, insisted Billroth, must come from a sound understanding of underlying pathological changes to the body such as would arise from inflammation or haemorrhage, for instance. Armed with such knowledge, Billroth and his students embarked on a programme that was, in the words of the historian Roy Porter, 'Frankly experimental, his new methods sacrificed many lives but, as his practices became refined and post-operative care improved, mortality rates dipped.'[24] For the students flocking to his school, Billroth seemed to be opening the way to a new era in which surgery could truly become the premier instrument of restoration and cure of the sick body. The rise of anaesthesia, as well as the introduction of antisepsis and aseptic techniques and then, in time, the ability to visualize the interior of the body via endoscopes and x-rays, only added to this sense of new frontierism. As Porter argues, the spirit of the age of colonization was not just confined to matters of geography: 'The body's interior seemed an Africa in microcosm, that dark continent opened up, mapped and transformed. Fame and fortune awaited the surgical pioneer who first laid the knife to some hitherto untouched part— perhaps he would be immortalized by an eponymous operation.'[25]

Soon after his arrival at Vienna, Billroth published another model text for the new academic surgery, *Chirurgische Klinik Zürich, 1860–1867*, a series of case studies drawing from his previous post in Zürich.[26] The cases contained in *Chirurgische* covered several remarkable techniques in general, abdominal, and cancer surgery, including a report of a complete prostatectomy. In that case Billroth diagnosed urinary obstruction in a man of thirty caused by a tumor of the prostate. He removed the gland using an extraurethral method (he would later switch to a transpubic approach) and reported that the prostatectomy was successful in that the patient recovered from the operation, although he did die of a recurrence of the tumour eighteen months later.[27]

The incidental removal of prostatic tissue during lithotomy was not uncommon and the literature before the nineteenth century contains references to the supposed subsequent benefits in affected patients.[28] Nor was Billroth the first to deliberately remove the whole prostate to treat a suspected cancer. In an 1885 review of the field of prostate surgery, the prominent New York surgeon, John Gouley, mentioned several reports (mostly from European texts) where controlled extractions of prostatic tissue had been attempted from the earliest decades of the nineteenth century.[29] The fame of Billroth and his school did serve to increase awareness of such surgeries, however, influencing especially those young American clinicians then travelling the great clinics of Europe. A newly minted surgeon by the name of William Halstead was one such traveller, and he brought a very Billrothian vision of surgery and its place in medicine back with him to the United States. Along with William Osler, William Welch, and Howard Kelly, Halstead would become one of the original 'big four' that helped to make the Germanic-style medical school at Johns Hopkins such an influential success. As mentioned above, Halstead would also become famous for the 'Halstead procedure' of radical mastectomy to treat breast cancer (an operation that would stay in vogue, little altered, until the mid-1970s when its drastic nature and disfiguring effects were called into question). Aside from cancer of the breast, Halstead experimented with radical surgery aimed at the treatment of other cancers too, and it was in this work that he inspired his junior colleague Hugh Young to follow in his footsteps by perfecting techniques for total resection of the prostate.[30]

RADICAL SURGERY AT JOHNS HOPKINS: HUGH YOUNG AND THE RISE OF ACADEMIC UROLOGY

In 1903 Young published the results of prostatectomies on some fifty patients from which he concluded that the procedure offered a cure for prostate cancer that showed no local spread.[31] This paper also introduced some new devices and techniques that would soon become ubiquitous in perineal prostatectomies including 'Young's urological table', for the optimum positioning of the patient, and a prostatic tractor to draw down and visualize the prostate prior to excision. In 1910 Young published another series of results, this time drawing from some four hundred patients, and establishing his technique as the model surgery for prostatectomies not least because of his impressively low mortality rate of just over three per cent.[32] By the early 1900s, prostate surgery, like many other surgeries, was much written about in the new professional journals then springing up as surgeons vied for the prestige of claiming the definitive technique. In this regard Young was also very successful, and the radical prostatectomy that he perfected bears his name to the present day.

Young regarded his operation as effective for benign hypertrophy (a cellular concept he took from Virchow) in addition to prostate cancer. Due to the size of his patient pool, Young was able to confirm some three or four decades after the musings of Wyss and Jolly that prostate cancer, while not as common as benign hypertrophy, *was* a common cancer in men over forty (he estimated that cancer occurred as about one-fourth as often as hypertrophy, which proved to be a significant overestimation as more became understood about the incidence of hypertrophy in aging men).[33] Based on his observations, Young made some important characterizations of prostate cancer that would go on to shape the emerging field of urology. Chief among these was his claim that cancers emerged from the posterior portion of the gland, could develop quickly or slowly, and spread outside the capsule via Denonvilliers' fascia to invade the bladder and surrounding structures.[34] He was also highly cognizant of the fact that the spread of cancer signalled a very poor diagnosis. In keeping with the consensus emerging at the time, Young felt confident in presenting prostate cancer as a disease that increased in incidence with age, so bringing to a conclusion several open questions circulating during the late nineteenth century. By the 1910s then, Young and Johns Hopkins were at the centre of prostate surgery as was emerging as part of the specialism of urology. As Young's fellow urologist Edward Lawrence Keyes apparently quipped, 'The prostate makes most men old but it made Hugh Young.'[35]

That Young was able to build a programme and reputation around a disease and an organ at the margins of surgical practice only a few decades before is a remarkable illustration of how rapidly the knowledge, institutions, and professional status of medicine were changing at the turn of the nineteenth century. Just as Halstead had done for his surgery of the breast,[36] Young pushed his method as one that could result in complete cure if doctors, especially general practitioners, could be taught to look for and understand the significance of early detection:

> [T]he early diagnosis of carcinoma is generally easy. A very hard nodule or area palpable by rectum, unless shown by x-ray to be calculus, should be looked upon with grave suspicion. In such a case the patient should be subjected to perineal operation at which the lesion is seen, palpated, and if necessary, an area incised and studied microscopically. When the diagnosis is confirmed, my radical operation can be performed with a prospect of radical cure in probably 75 per cent of the cases. If general practitioners could be taught to make rectal examinations much more frequently, and be suspicious of every markedly indurated prostate, even when only a small nodule is present, many patients would be brought to early radical operation and ultimate cure.[37]

The academic context in which Young developed his teaching and experimental programme is crucial to an understanding of its wider significance. The French created one of the earliest professional urological associations when they founded *l'Association Française d'Urologie* in 1896, an organization that inspired the creation of the

American Urological Association (AUA) by Young and others in 1902 (a German association followed in 1908, and at a meeting in Paris in 1910 an International Urological Association was founded, the *Société Internationale d'Urologie*). The US did, in fact, have an organization that preceded the founding of the AUA—the American Association of Genitourinary Surgeons was set up by Keyes and colleagues in 1886—but that group confined themselves almost exclusively in their early years at least to venereal disease (although this did involve considerable discussion of the prostate as a common consequence of gonorrhoea was prostatic inflammation, or prostatitis, that some thought required removal of the gland). The AUA by contrast was much more broad and ambitious in its scope—perhaps particularly in its encouragement of physiological experimentation on animals—but it seems that the organizations existed side-by-side peaceably, and several prominent surgeons, Keyes and Young included, held membership of both.[38] It is telling, however, that a commentator (one Ernest G. Mark of Kansas City) at the AUA meeting of 1911 noted, 'I believe we men who are making a specialty of urology should discountenance the old idea of "clap doctor." That is where urology first got its black eye.'[39] The Chicago surgeon G. Frank Lydston agreed that the association with venereal disease had left the urologist as, 'an appendage and handmaiden to general surgery', and the competition from general surgeons made it all but impossible for a urologist to build up an independent practice.[40]

Part of the problem in retelling history as a story of great men and great advances is that we can forget to ask some basic questions like, how common was this or that operation? How much, if at all, did any single advance or operation change the lives of men with prostate cancer? As an historical resource, the records of professional organizations can provide an invaluable glimpse behind the veil of optimism. It seems obvious to say that techniques like Young's were mostly confined to the few specialist urological facilities dotted around the United States, but gauging how widespread the knowledge and procedures of radical prostate surgery actually were is a difficult task unless the historian has access to some frank and unguarded discussion. The 1908 edition of *Transactions of the American Urological Association* covering the seventh annual meeting of the Association in Chicago, for instance, provides an excellent illustration of what such sources can say about general professional attitudes.

In a piece called Confessions of a Yeoman Prostatectomist an Indianapolis surgeon Joseph Rilus Eastman was forthright about his admiration for the urological 'Greats':

> As we stand marveling at the machine-like technic of Young and Mayo and Freyer, among other impressions which we chronicle is the one that the methods which bear the names of these men are theirs, indeed, and no one's else in the same sense. Their records pertain less to their methods than to themselves and their genius.
>
> ...
>
> Each has amazingly good results, yet each condemns the other's plan. ... The man is more important than the methods.

Eastman roundly condemned the bad work of inexperienced surgeons who he considered were dragging the procedure into disrepute with their high mortality rates and damage to sexual and urinary functions and then blaming the technique and not themselves for the failures. That said, he also vented his own frustration with trying to emulate the great urologists stating:

> Like many others, I have found it difficult before operation to distinguish between benign and malignant enlargements of the prostate gland. ... There is no doubt that some men of great experience have the acumen necessary to determine quite definitely before operation the character of the prostatic enlargement, a faculty quite beyond me and many of my colleagues...[41]

As Eastman dryly noted, prostate surgery was rarely carried out on patients in any kind of healthy state: age and existing damage to the urinary system from retention and repeated catheterizations all took their toll on mortality figures; as Eastman put it: 'Surgery cannot rectify in as many minutes conditions which have taken years to obtain.'[42] While Eastman did not doubt the power of surgery under the best conditions, he was adamant that few doctors actually practised in the best conditions. Speaking of injuries that an inexperienced operator can make to the ejaculatory ducts, sphincters, and rectal wall, Eastman said of the latter:

> It is my observation that injury to the rectum occurs ... because of too great lateral traction by too powerful pulling on the retracters, as the result of which the rectum is split. [Experts say] none but an experienced and trustworthy assistant should be allowed to hold the retracters. We who are

obliged to operate in country homes and are compelled by professional ethics to allow the family physician to assist, have usually the greatest sense of
security when such an assistant is placed at the distal end of a long-handled
retracter, but in a prostatectomy he is by no means innocuous in this relation, as I have found to my sorrow. Such an accident would never occur in
a modern operating room with a well-trained operating corps, but I would
take my oath that they do occur in farmhouses in Indiana.[43]

In spite of the hardships, Eastman recommended the technique and,
following Young, urged for early intervention and better education of
the profession at large: 'When the profession generally appreciates the
importance of early prostatectomy before serious complications have
developed, then the prostatectomist of average skill, the yeoman, may be
able to approach more nearly to the standards of the masters.'[44]

In the discussion that followed, the once again sceptical Lydston
underscored the sharp divide between the 'masters', the 'novices', and the
'yeomen':

With reference to the encouragement that has been offered the novice by
publishing reports of prostatectomies and brilliant descriptions and pictures
of operations, I have a decided opinion which I am going to express without any intent to be discourteous or to discredit any man or his work. I am
going to take as a type that has done the most damage, the operation of
our distinguished President, Dr. Young, and the operation of Park Syms of
New York. I have in my mind's eye a very vivid recollection of these operations as they appeared in the Journal of the American Medical Association
and elsewhere. A very beautiful and attractive operation, an operation that
would attract the novice and convince him that all he had to do was to make
a median incision in the perineum, with a retractor pull the prostate down,
pry it out of the perineum, make an incision first on one side and then on
the other, and fish out the lobes. I suspect that the gentlemen when they
published these descriptions of their operations intended to be diagram-
matic and rather conventional in the matter of illustrations. If so, however,
they failed to so state.[45]

In response to these criticisms, Young first defended his own successful
record with the operation (in addition to his other comments, Eastman
had queried whether the reports coming out of Hopkins were as transparent as they might be) and then protested that the operation was, indeed,
safe for the average surgeon, at least one that knew his anatomy[46]:

I cannot understand why people have so much trouble with what really is one of the simplest operations that I do. It is apparently really no trouble at all, and I do not honestly see how anybody can have trouble with it. And when they do have trouble they had better do something else. I do not urge it. If anybody will come for a month to Baltimore I will try to show him twelve or fifteen prostates, and if I cannot prove to him that I do not injure the ejaculatory ducts or the internal sphincter, I will give him a big terrapin dinner.[47]

Young's comments failed to mollify the critics and the St. Louis surgeon, Bransford Lewis, urged Young to take their difficulties more seriously: 'It is not right to portray this operation as being unvaryingly successful and simple. It is not, absolutely not.'[48]

Another aspect of Young's reasoning called into question at the meeting was his advocacy of prostatectomy as a cure for prostate cancer. Regardless of the difficulty (or not) of the operation, several surgeons were sceptical of the notion of 'cure', feeling that reappearance was to be expected especially since, as Keyes remarked, 'I am extremely conservative in operating for carcinoma of the prostate. ...[T]hese old gentlemen commonly have no symptoms indicative of carcinoma until it has exceeded the limits of the prostate, and has become to all intents and purposes inoperable.'[49] Young himself had, in fact, by 1908 retreated from his earlier ebullience regarding cure, stating at the annual meeting that he believed 'most' cases that come to the surgeon are too far advanced for cure.[50] He did, though, point to the palliative benefits of the operation:

There are ... cases which I think can be operated upon and given relief, although they cannot be cured of their cancer. I have had now about 10 cases in which, although I knew they had cancer and that a radical operation would not cure them, I have simply gone in and removed the obstructing parts of the cancer around the urethra and given the patient comfort—one case for two years, another three, and another almost a year.[51]

A major problem for Young's fellow surgeons, though, was seeing prostate cancer as developing in a neat, linear progression. Other surgeons present recorded their observations that when supposedly benign hyperplastic tumours were removed and examined under the microscope carcinomas were routinely found, further adding to the confusion over the status and clinical course of prostatic cancer.[52] Similarly, another feature of Keyes conservatism with his scalpel was, he explained, his belief that some tumours behaved in ways that did not seem to justify the risks of intervention, '[W] hen one does find a case of carcinoma of the prostate in which symptoms occur before the carcinoma has exceeded the limits of the gland, I have been surprised to find how extremely slowly the growth progresses.'[53]

Beyond questions of pathophysiology of the prostate though, lay more fundamental issues about the status of urology as a specialty at all. In his Presidential address to the 1911 meeting of the AUA the founder of the urological service at the Massachusetts General Hospital, Hugh Cabot (younger cousin to Arthur Tracy Cabot), spoke to the strength of urology's claim:

> The claim of urology to be regarded as a specialty began with the development of intricate and difficult methods of diagnosis and treatment. The advent of the cystoscope, the ureter catheter, the development of intravesical operations and the application of the X-ray to diagnosis and conditions of the urinary tract, have taken it out of the field of the general practitioner.[54]

It was not just that the mastery of urology was beyond the ken of the general practitioner; Cabot argued that the frequency of errors in diagnosis and injury through mistakes in treatment made the urological patient poorly served by the general surgeon. This was an important distinction, he argued, since few surgeons wishing to confine their practice to urology were likely—as might be expected by a successful general surgeon—to secure control over beds in large hospitals, and were instead confined to the less prestigious outpatient clinics. Cabot went on to list the various diseases that he regarded as requiring the attention of a specialist, saying of prostate problems:

> In the last fifteen years the diagnosis and treatment of obstructing lesions of the prostate have been entirely revolutionized. Previous to that time surgical treatment was unsatisfactory and the operative mortality so high as to be almost prohibitive.
>
> Under the combined efforts of a few genito-urinary surgeons in this country and abroad, the diagnosis has become accurate, the indications for operations have been made clear and the technique of prostatectomy, both suprapubic and perineal, has been perfected.
>
> Let it be observed ... that this revolution has been due to the work of men essentially specialists. At the present time the operative mortality in the hands of skilled operators has been reduced below six per cent. The proportion of radical cures is very high and the number of cases unimproved, or

worse, small. With this standard of work of the general surgeon will not compare favourably. Statistics are difficult to collect, but efforts made by White of Portland, and Moore of Indianapolis, tend to show that the general mortality in these cases is still above fifteen per cent.

In many of the large municipal clinics today there frequently appear cases in which prostatectomy has left the patient far worse than before, owing to damage to the rectum, failure of the urinary fistula to close or a joint urethra-rectal fistula. Such accidents are rare in the hands of a specialist, common in the hands of a general surgeon.[55]

Cabot further concluded that patient management was not yet in step with the scientific management of diseases: 'In the management of diseases of the prostate, therefore, our knowledge has advanced rapidly and our results can be made good, but the patient is not today receiving, in many instances, the benefit of the knowledge that exists.'[56] He further noted the reluctance of general surgeons to refer cases of stone in the bladder and diseases of the kidney. In the comments that followed Cabot's address, however, we see a vigorous push back based on the perception of the general surgeon. J. Bentley Squier noted:

It is my judgment that a great many men who have gone into urology have not had the training in the early part of their career to make them competent to undertake major operations. In any large city we can pick out a number of men who are adequately equipped, but the majority of urologists are not up to it, they have their training in dispensaries, and have not had general surgical appointments. ... The future of urology depends upon educating the urologist to first have a thorough training in general surgery and then devoting himself to his specialty.[57]

Something of a Catch-22, in other words, and one that led him to pessimistically conclude, 'I do not believe we will reach that beautiful heaven which our chairman would like to have us in.'[58] Keyes, however, resisted such an argument saying:

Let us work for the advance of our art with the absolute assurance that if we do so we shall at least earn a living wage, that we shall see all over the country an ever-increasing multiplication of hospital genito-urinary departments, that our hearts shall be cheered with the realization that those better equipped men who will come after us will reap the fruits of our labors, and that nothing under God's blue heaven can prevent us from being specialists.[59]

In spite of his bold words, Keyes' comments did point to the underlying problem of institutional and economic opportunity for would-be specialists. While Young's urological practice at Hopkins had greatly benefitted from a large gift from the railroad tycoon and famous gourmand James 'Diamond Jim' Buchanan Brady (who Young had treated for prostate problems), few devoted departments of urology existed elsewhere in the States at that time.

It seems likely that ongoing scepticism of the methods of prostatectomy combined with limited professional opportunities probably limited routine surgical intervention on the prostate in the first two decades of the twentieth century. Nonetheless, the prostate remained of considerable interest to surgeons and continued to be written about. The University of Pennsylvania's professor of surgery, John Deaver, for instance, wrote an entire textbook on the normal and abnormal prostate in 1905 called *Enlargement of the Prostate: Its History, Anatomy, Etiology, Pathology, Clinical Causes, Symptoms, Diagnosis, Prognosis, Treatment, Technique of Operations, and After-Treatment* that proved popular enough to pass into a second edition in 1922.[60]

By the mid-1930s the turn towards specialty practice in American medicine was a rising tide that lifted all boats, including that of urology (if slowly). Take the career of Frank Hinman, for example, a graduate of Johns Hopkins and a trainee of Young's. Like others of Young's students, Hinman went on to establish his own academic department of urology, and pushed for the recognition of urology as a specialty. If any specialty was to survive, Hinman realized, it needed practitioners who had access to hospital beds and training but who also had a means of making money in private office-based practice. In this latter regard, Hinman outlined 'the eight steps to a presumptive diagnosis'—patient history, physical examination, abdominal examination, external genital examination, urinalysis, prostate examination, x-ray examination, and a PSP (phenolsulfonphthalein) test for kidney function—that would allow the urologists make diagnoses non-invasively and in an office setting.[61] Hinman's *The Principles and Practice of Urology* published in 1935 was a text for the student for sure, but also, and importantly, a guide to the general practitioner so that they might recognize a urological case when they see one and refer the patient on appropriately. The book sold poorly,[62] but it was an interesting early model of how a specialty textbook could target an audience beyond a small pool of fellow enthusiasts and interested general surgeons.

Another aspect of Hinman's notable career was his role as one of the experts called by the AMA to establish the American Board for Urology in 1933.[63] Like the other boards established by the AMA during the interwar years, the new organization sought to regulate the practice of urology by defining postgraduate, residency training routes and imposing stringent examinations which the would-be specialist must pass to be 'certified'. The AMA also lobbied the individual states to accept the new standards as the states were, as they still are now, responsible for the licensing of doctors. The AMA hoped to convince state licensing boards to deny and exclude practitioners who did not meet the new standards of education and training, or who exceeded a certain scope of practice and encroached upon specialisms without the requisite credentials. The process was not smooth, but the AMA (aided by other professional groups such as the American Hospital Association) was ultimately victorious in forcing regulation of medical practice through academic reform, and urology, like the other specialties, became formally recognized and formally defined within this process. As the historian George Weisz put it:

> It is difficult to overstate the significance that [the American system of certification] brought about. In the space of several decades, the majority of specialists were transformed from generalists who gradually took on specialist work in the course of their careers to exclusive experts whose specialty training in hospital residency programs, followed by certification examinations, was the culmination of their medical education. In the context of this new form of training, part-time specialist work made very little sense, even when it was not expressly forbidden by a specialty board. Consequently, part-time specialists, who made up more than half of all specialists in 1925 and 1930, became increasingly rare; by 1945 they comprised a tiny proportion of the medical population.[64]

This move toward specialization was illustrative of larger trends in American medicine that saw the United States come to dominate medical research and training in the latter part of the twentieth century. In the period following WWII, massive federal investment in the National Institutes of Health (NIH) combined with expanding academic, pharmaceutical, and device-manufacturing sectors, helped the US push aside the older European centres of excellence to become a world leader of unprecedented resource and ambition. Where once US doctors had flocked across the Atlantic from west to east, by the 1950s and 1960s that flow was quite reversed. For sure the traditional centres of excellence—London, Paris,

Berlin, Vienna—remained excellent and continued to innovate (the retropubic prostatectomy, an operation that joined Young's perineal operation in becoming standard in modern surgery, was developed in London by the Irish surgeon Terence Millin, for instance[65]) but the US became the frame of reference, the one to beat, or, more usually, imitate. In the chapters that follow I will mostly confine the continuing story of prostate cancer to US cases because, for good and for bad, we most clearly see there the pathway of a common disease coevolve within the modern biomedical complex.

Therapeutic cutting is as old as human society and while healers differed in their views on if and when the knife was appropriate, instruments and operations to relieve urological stones and blockages were some of the most ubiquitous historically speaking. The long nineteenth century was a time of unprecedented change for the practice of urological surgery as the ancient practices of lithotomy and catheterization inspired new aggressive interventions in disease. The decline of humouralism and the advent of a new materialism in pathogenesis created a new rationale for these interventions, just as better controls of pain, bleeding, and infection made their uptake more practicable. For cancer, a disease that Galenism had long advocated as suited for the knife, the new surgery seemed to offer the possibility of cure, a sentiment perhaps most strongly expressed in the work of the Johns Hopkins surgeons, Halstead and Young. Others were not so convinced, however, and it would soon become clear even to Young that the surgical control of prostate cancer was something of a chimera. The likelihood of finding cancers before they had spread outside the prostate capsule and begun to cause clinical symptoms was not high, especially since the use of digital rectal examination did not seem to be a part of routine physical examinations in the offices of general practitioners until a much later period. Notwithstanding this rise and fall of surgical optimism for prostate cancer control, the 1930s did see the development of some far-reaching ideas in hormone therapy as the role of hormones in the pathophysiology of cancer became better understood. It is to these innovations that I will now turn.

NOTES

1. Cooper and Tyrrell, *The Lectures of Sir Astley Cooper*, 223.
2. Young, Radical Cure of Carcinoma of the Prostate.
3. Desault, *Traité des Maladies des Voies Urinaires.*
4. Bynum, *Science and the Practice of Medicine in the Nineteenth Century*, 9.

5. Porter, *The Greatest Benefit to Mankind*, 530–2.
6. Starr, *The Social Transformation of American Medicine*, 90–1.
7. On opposition to specialization from within the medical profession see Harley Warner, From Specificity to Universalism in Medical Therapeutics.
8. Hippocrates et al., *Hippocratic Writings*, 67.
9. Riches, The History of Lithotomy and Lithotrity.
10. Ibid., 193.
11. Murphy and Desnos, *The History of Urology*, 152.
12. Ibid., 180–1.
13. Hunter, *A Treatise on the Venereal Disease*, 110.
14. Hunter, *Observations on Certain Parts of the Animal Oeconomy*, 38.
15. Ibid., 39. There is some debate about who was actually first to describe the gland that came to be known as 'Cowper's gland', but the London surgeon and anatomist William Cowper first discussed it in his own work in 1699. See, Sanders, William Cowper and His Decorated Copperplate Initials, 7.
16. Hunter, *Observations on Certain Parts of the Animal Oeconomy*, 33.
17. 'Lectures on the Principles of Surgery', 8.
18. Ibid., 9.
19. Mercier, *Recherches Sur Le Traitement Des Maladies Des Organes Urinaires*, 118, 127–9.
20. White, I. The Results of Double Castration in Hypertrophy of the Prostate, 58.
21. Cabot, II. The Question of Castration for Enlarged Prostate.
22. Porter, *The Greatest Benefit to Mankind*, 599.
23. Billroth, *General Surgical Pathology and Therapeutics*, 1.
24. Porter, *The Greatest Benefit to Mankind*, 599.
25. Ibid.
26. Billroth Chirurgische Klinik, Zürich, 1860–1867.
27. Ibid., 548–9.
28. Ricketts, *Surgery of the Prostate*, 8.
29. Gouley, Some Points in the Surgery of the Hypertrophied Prostate, 179–81.
30. For a discussion of the concepts of 'conservative' and 'radical' forms of surgery in the late nineteenth century, see Brieger, From Conservative to Radical Surgery in Late Nineteenth Century America.
31. Young, Radical Cure of Carcinoma of the Prostate, 32.
32. Young, Perineal Prostatectomy, 791.
33. Ibid., 785.
34. Young, Radical Cure of Carcinoma of the Prostate, 42. Denonvilliers' fascia is the fibromuscular structure beneath the prostate that separates the prostate and bladder from the rectum named for the French anatomist and surgeon Charles-Pierre Denonvilliers.

35. Murphy and Desnos, *The History of Urology*, 396.
36. For an account of the early twentieth century debates regarding 'delay' in the treatment of breast cancer, and for a discussion of Halstead's role in these, see Aronowitz, Do Not Delay.
37. Young, Radical Cure of Carcinoma of the Prostate, 46.
38. Lewis, The Dawn and Development of Urology, 17–20.
39. Cumston, *Transactions of the American Urological Association*, 14.
40. Ibid., 15.
41. Eastman, Confessions of a Yeoman Prostatectomist, 143.
42. Ibid., 146.
43. Ibid., 151.
44. Ibid., 152.
45. Cumston, *Transactions of the American Urological Association*, 159.
46. Ibid., 163.
47. Ibid., 164.
48. Ibid.
49. Ibid., 165.
50. Ibid., 172.
51. Ibid., 172–3.
52. Ibid., 168.
53. Ibid., 166.
54. Cabot, Is Urology Entitled to Be Regarded as a Specialty?, 2.
55. Ibid., 3.
56. Ibid., 4.
57. Cumston, *Transactions of the American Urological Assocation*, 10–11.
58. Ibid., 18.
59. Ibid., 20.
60. Deaver, *Enlargement of the Prostate*.
61. Bloom and Hinman Jr, Frank Hinman, Sr, 877.
62. Ibid., 878.
63. Ibid.
64. Weisz, *Divide and Conquer*, 144.
65. Millin, Retropubic Prostatectomy: A New Extravesical Technique.

CHAPTER 4

Sex, Hormones, and Quantification

The sexual organism, of which the prostate is one of the chief factors, is so inti-mately blended with the central and sympathetic nervous systems, that disease of this gland provokes the most varied neurotic disturbances. … Often I have seen men who had been dosing their stomachs for dyspepsia, their livers for tor-por, their bowels for constipation, their heads for neuralgia, treating sciatica for malaria, plastering their backs for Bright's disease, taking sea voyages for melancholia, when the origin of their trouble was centered on the prostate, and the relief of which cured their other ailments. George Whitfield Overall, *A Non-Surgical Treatise on Diseases of the Prostate Gland and Adnexa* (1906).[1]

One of the directions that cancer research is now taking is the functional or physiological approach to the problem of tumors. The functional approach contrasts sharply with the descriptive approach—with the methods of classical pathology. It is concerned with the entire living organism rather than with sections or segments of the dead organism. In the functional approach the mea-sure is of first importance: How much cancer activity is present? How can the activity be increased or decreased? Assay of a disease in a laboratory obviously removes much of the uncertainty inevitably associated with bedside observa-tion, particularly in cancer. The yardstick in prostatic cancer concerns certain enzymes, the phosphatases. Charles Huggins, Endocrine Control of Prostatic Cancer (1943).[2]

The revival of microscopy, the development of cell theory, and the emergence of specialization all drastically reshaped medicine during the latter nineteenth century, but so too did the development of another phenomenon I have not yet discussed: the science of physiology. Once

© The Editor(s) (if applicable) and The Author(s) 2016   69
H. Valier, *A History of Prostate Cancer*,
DOI 10.1057/978-1-137-56595-2_4

again the great medical centres of Europe—London, Paris, Berlin—were at the epicentre of profound transformations in the professional, practical, and intellectual development of medicine and surgery, and nowhere was this more apparent than in the Paris laboratories of Claude Bernard. Bernard's classic *Introduction à l'Étude de la Médecine Expérimentale* (An Introduction to the Study of Experimental Medicine) was published at the peak of his career in 1865, by which time the Emperor himself, Napoleon III, had promised him a new laboratory suite at the *Muséum National d'Histoire Naturelle* in Paris. Bernard was an aggressive and ambitious experimenter, freed from the necessity of supporting himself through the practise of medicine thanks to the financial support of his wife, Marie Françoise Martin, and the professional support of an early mentor at the *Hôtel-Dieu*, the physician and physiologist, François Magendie, who provided Bernard with early access to laboratory equipment. A strict empiricist, Bernard proposed what would become the classic experimental model in science: determination of cause and effect not through correlation but through the control and manipulation of variables. If variables could be altered independently of each other, and reliably caused an observable effect to occur then, and only then, he reasoned, could a causal relationship be established. For Bernard, observation of the sick patient at the bedside, even when supplemented by newer tools, like the stethoscope, limited the science of medicine to a descriptive practice. Similarly, the lesions and diseased organs for so long the focus of attention by morbid anatomists represented the endpoints and not the dynamic *process* of disease itself. In place of such 'passivity' then, he proposed a continuation of Magendie's interventionist, animal-based (vivisection), approach to the investigation of normal and abnormal physiological function under the most strictly controlled laboratory conditions. In his preface to *Médecine Expérimentale*, Bernard was clear about the relative benefits of these methods of investigation:

> In the empirical period of medicine, which must doubtless still be greatly prolonged, physiology and therapeutics could advance separately; for as neither of them was well established, they were not called upon mutually to support each other in medical practice. But this cannot be so when medicine becomes scientific: it must then be founded on physiology.[3]

For Bernard, there could be no rational therapeutics, no real pathological understanding, without the efforts of the physiologist.

Beyond his contributions to physiology, Bernard was an active spokesperson for the new 'biomedical' model within the institutions of medicine. He abhorred the fact that experimentalists, deprived of space and salary, often struggled to make careers in science. Income from practising medicine part-time enabled independent investigators to support themselves to some extent, but the sheer scope and complexity of the new laboratory sciences was fast making this kind of *ad hocery* unfeasible for the serious scholar. While sister sciences in bacteriology, synthetic pharmacology, and even immunology (via vaccines and anti-toxins) had clear connections to industry and public health, this was not so (or at least not apparently so), with the more abstract research of the physiologist. For biomedicine to flourish in the way that it did during the twentieth century would require a transformation of the institutional and economic base of academic careers at least as far reaching as the professionalization and specialization of medicine described in the previous chapter.

This new physiological bent in medicine also helped to fundamentally change our understanding of diseases of the prostate, including cancer, through the science of endocrinology. Glandular (spongy) structures had been described as far back as Galen, and morbid anatomists from the Renaissance onwards had speculated that glands might release secretions into the blood, but it was a physiological model of the body that placed endocrinology at the forefront of biomedical research at the turn of the nineteenth century. Bernard's proposal of a *milieu interior*—a dynamic balance—raised questions about the role of these secretions in the regulation of all the functions of the body, from heating and cooling, to the metabolism and absorption of food. His vivisections, furthermore, showed that scientists could begin to answer these questions by experimentally inducing crises arising from the removal of certain ducts and glands. One devotee of Bernard's approach was his successor at the Collège de France, Charles-Édouard Brown-Séquard who took up the chair of experimental medicine following Bernard's death in 1878.

Brown-Séquard was a master of experimental manipulation of glands through vivisection (and was particularly well known for his work on the adrenal glands in this regard), and he, along with his mentor Bernard, made numerous discoveries concerning the nervous system of the spinal cord. In addition to these laboratory researchers, Brown-Séquard gained considerable fame for his efforts towards the application of physiological know-how to therapeutics. Like Bernard, Brown-Séquard was convinced that the *milieu interior* required constant regulation from some kind of biological messenger emanating from the organs and glands, but he also believed that these messengers could be harnessed for therapeutic purposes. In 1889 he made a presentation to the Société de Biologie (an organization that Bernard had helped found in Paris two decades earlier) announcing his self-experimentation with *liquid testiculaire* injections made from the macerated gonads of dogs and guinea pigs.[4] He believed that his preparations could serve to 'rejuvenate' the aging male, improving everything from mental capacity to sexual vigour.

The connections drawn by Brown-Séquard between spermatic production, the testicles, and the physical and psychological state of the male were in many ways as old as dirt. The use of castration in animal husbandry to calm and tame animals dates back to Neolithic times, after all. Similarly, in the writings of the first century CE Roman naturalist Pliny the Elder we see the advice that aging men who care to maintain their health and vigour ought to feast on animal testicles. Closer to the time of Brown-Séquard was the work of Arnold Adolph Berthold who made the experimental observation that castrated roosters—capons—could be (somewhat) restored to their former virility when implanted with donor testicles (in this case inserted within the abdomen).[5] In a related sense, Brown-Séquard himself noted that it was a medical 'fact' 'long known' to all practitioners that 'spermatic anemia' could pose a serious threat to the health of the male:

> [T]he mind and body of men (especially before the spermatic glands have acquired their full power, or when that power is declining in consequence of advanced age) are affected by sexual abuse or by masturbation. ... [I]t is well known that seminal losses, arising from any cause, produce a mental and physical debility which is in proportion to their frequency.[6]

Indeed, the pathological role of spermatic loss in sexually active men was of considerable concern to the urologists at the turn of the nineteenth century, many of whom connected problems with the prostate with the habit of making second marriages later in life.[7]

SEX, GLANDS, AND MASCULINITY

George Whitfield Overall's *A Non-Surgical Treatise on Diseases of the Prostate Gland and Adnexa* quoted above was one of the first (of many) attempts by medical writers to promote the idea that normal masculinity rested in the normal functioning of male sexual parts.[8] In Overall's case, he sought to educate general practitioners about the centrality of the prostate to a man's overall health and to advise other specialists about standard treatment options. The role of venereal (and other) infections, benign hyperplasia and cancer in causing urinary discomfort and other tell-tale symptoms of a diseased prostate I have discussed in other chapters. What is different here is that Overall explicitly connected the disordered prostate in particular to the kinds of psycho-sexual dysfunctions referred to above. The prostate, he reasoned, was particularly prone to damage given the fine line between 'normal' sexuality and 'excessive sexual indulgence':

> During erotic excitement, whether normally or abnormally, the prostate becomes hyperemic, either synchronously with or independent of penile erection. If this excitement is unduly prolonged, by toying with women, indulging continuously in libidinous thoughts, association with prostitutes, masturbation, continence or excessive intercourse, it causes venous stasis or congestion of the gland resulting ultimately in subacute or chronic prostatitis; which readily extends and involves the prostatic urethra and adjacent parts.[9]

As his *Treatise* ran to several editions this was clearly advice that other physicians felt they needed to hear.

The assumed relationship between sexual excess and physical, spiritual, and mental decay has deep historical roots. Within the Victorian fever for moral reform, restraint was key, particularly since the modern world with its incredible transformations in population growth, mass immigration, and industrialization also brought with it panics over crime, misery, disease, and filth. Nowhere is this better exemplified than in the disease category of 'neurasthenia', or nervous exhaustion.[10] Rapid urbanization and the increasingly frenetic pace of life seemed to be tearing at the fabric of older ways of being, and the mind, the nerves, were vulnerable. While neurasthenia was a category almost exclusively tied to the psyche of men, women too were considered vulnerable to 'nervous illnesses' although in their case this was usually regarded as 'hysteria'.[11] While the fractious stimuli of the external world were regarded as a major cause of such mental collapse, the behaviour and biology of the individual also played their part. While medical writers pondered the role of the womb in 'hysteria', so others wondered about the correspondences between the uterus and the prostate. As Overall said of this in a section of his treatise devoted to neurasthenia:

> Genito-urinary diseases of men as [a] result of prostatitis and the various functional nervous disorders related thereto, whether as cause or effect, are in the same condition that diseases of women were in fifty years ago. At that time the nervous symptoms that accompanied such disorders in females as lacerations of the cervix or perineum, congestion and displacement of the uterus and ovaries, were succinctly, if unscientifically, grouped under the head of hysteria, and these symptoms treated without reference to the cause and often without the least effort to arrive at a correct diagnosis. And today the nervous maladies resulting from a morbid condition of the prostate gland, such as mental depression, morbid fears, nervous dyspepsia, palpitation, deficient mental control, headache and backache, are generally dismissed in the same easy fashion to the category of hypochondriasis.
>
> Considering the immense importance of the problem involved in the relation of the genital function to the nervous system, and the vast amount of suffering entailed upon mankind by the ignorance of the patient and the indifference of the physician in regard to these problems, remarkably little effort has been expended in their solution.[12]

Pathology in the prostate was then (according to Overall at least) to be suspected in all cases of nervous troubles in men, who, just like women, were apparently liable to failures in health due to their seemingly disaster-prone sex organs.

The idea that 'respectable' men would engage in non-procreative sexuality seemed to be both a reflection and a further threat against an ostensibly fragile *fin de siècle* social order. Take masturbation, specifically: while social and religious prohibitions against the activity date back to ancient times, by the nineteenth century the pathologization of the practice was in full bloom. The anti-masturbation crusades of the period were waged with particular vigour in the United States. Benjamin Rush's highly influential publication *Medical Inquiries And Observations Upon The Diseases Of The Mind* (1812) is a prime example of how 'onanism' was linked to mental disease and physical ailment in the minds of the American medical elites. Other crusaders were keen to promote the view that masturbation was perilous not only to the body and mind of the individual but to the social body as a whole.[13] The historian Frederick Hodges gives the example of the prominent Massachusetts surgeon Alfred Hitchcock, who, after gaining a stellar reputation for his actions during the Civil War, would go on to become a high profile advocate of anti-masturbation theories and the moral obligation of the physician:

There is a great reluctance on the part of the profession to 'speak the whole truth' on this disgusting subject. Does not this silence cherish the ignorance and weak prejudices of the community, and thus indirectly afford encouragement and patronage to boasting empirics and unprincipled medical pretenders. Shall we shut our mouths from candidly, and in a proper manner, speaking the truth to our patients, for fear of offending the pride of families? Shall we indeed, for selfish reasons, compromise the lives of our patients at the shrine of popular prejudice? Can we discharge our whole duty as laborers for the best good of suffering humanity ... and suffer the community to remain ignorant of this destroying Moloch of civilized society? Our profession, as a general thing, have nobly come forward and denounced intemperance as one of the greatest individual and national evils. ... But not so plain are the symptoms of the evil in question. It is insidious, but certain in its operation. Its course is silent and solitary, but mighty and ruthless are its movements. It steals unseen and almost unfelt, but blights and destroys like the breath of the sirocco. The manly frame totters and decays beneath its undermining power, while the social, moral and intellectual man is wrecked or annihilated in the ruin.[14]

The upheavals of urbanization and industrialization in postbellum America, Hitchcock warned, were leading to social dislocation as young men poured into the cities from their farms. As these youngsters turned to masturbation instead of marriage, so they engaged in a practice that would weaken and perhaps even kill them so undermining the stability, safety, and productivity of the new social order. If educated and professional men (who presumably should have known better) were to also turn to masturbation as a means to deal with social pressures, the future of America looked bleak indeed. As the famous neurologist George Miller Beard had argued in his hugely influential 1881 text *American Nervousness* (a book that popularized the very term 'neurasthenia'), the more civilized the man the more removed he was from the rougher constitutions seen in white working men and black men (of any social standing). Beard, in keeping with many other commentators in the racial science debates of the nineteenth century, believed the 'negro' to be closer to an animalistic state of nature than his white counterparts, and so immune to neurasthenia and impotence due to his 'natural' hyper-potent state. To a lesser degree, the white working class 'muscle man' shared this constitutional resistance to

the psychologically destructive effects of civilization but not the insus-
ceptibility against sexual degeneracy supposed of his already 'degenerate',
'barbaric' black brethren.[15] As the historian Kevin Mumford points out,
the stigmatization of masturbation in nineteenth century America served
an ulterior purpose of furthering the racial and social divisions of American
society. 'Impotence was a white man's problem,' Mumford argues, 'unlike
black men (and Victorian women), white men possessed the capacity to
exercise enough rational self-control to avoid the disorder.'[16] Scientists
and social reformers helped, in other words, to construct self-restraint
as the means for some men to not only avoid sexual disorder but to set
themselves apart as middle-class and white.

By the mid-nineteenth century, then, 'excessive' masturbation (whatever that meant, according to Overall opinions were mixed) was widely
regarded as but the first step to a state of serious disease. Masturbation-
induced spermatorrhea (a term that drew from John Hunter's description
of seminal weakness, *Treatise on Venereal Disease* to mean the discharge of
semen without an erection[17]), so the theory went, was the underlying cause
a of a wide variety of alarming disorders including neurasthenia, impotence, azoospermia, insanity, and, as we have seen, was supposedly responsible for premature death.[18] For societies concerned over declining birth
rates amongst the upper and middle classes, such wonton self-destruction
was anathema. In a stern notice to his fellow clinicians, the New York
Bellevue Hospital surgeon and professor Joseph Howe described how to
spot the chronic male masturbator:

> It is only when the sin has been besetting one for a long period of years,
> and when it has destroyed many of the finer instincts of manhood, that you
> notice the characteristic pale expressionless face, with sunken eyes, that sel-
> dom meet yours, but steal sideways glances, when your attention is drawn in
> another direction. Only in such chronic cases do we find the patients cow-
> ardly, easily startled, sleepless, inanimate, forgetful, stupid, troubled with
> vertigo and epilepsy mere animals in everything—in desire, as well as in
> action.[19]

Howe went on to share his belief that the observant doctor would find
other signs of sexual deviance in his patients through careful examination of the penis (liable to shrinkage and deformity), scrotum (often dis-
tended), and the urethra and prostate gland (likely to be heavily congested
with mucus, with the gland itself swollen and tender to the touch).[20] As

in other cases of prostatitis, then, treatments often involved the insertion of bougies and catheters although as a deterrent against future abuse and to make masturbation less pleasurable the surgeon might also have added cauterization or caustic or electroshock treatments to the unfortunate patient.[21]

With the medical profession beginning to pervade even 'normal' marital sex, through interdictions against delayed gratification, and too much, or too little, sex, so the pathologization of other expressions of male sexuality (homosexuality, extramarital promiscuity, and so forth) intensified. Conversely, the centrality of the role of the testes and prostate in the general mental and physical health of men was a source of major concern for urologists who, through treatment of cancer or prostatic enlargement, might disrupt the essential spermatic economy. Within the castration controversy surrounding the procedures of White and Cabot discussed in the previous chapter, for instance, we see how these considerations affected attitudes towards surgical intervention. Cabot, for his part, believed that White made too little of the effect of castration because he made too little of the testicles themselves and their role in making men healthy:

> by the removal of the testes, the vital force of the patient has been in some way diminished, and thus, in a measure, the theory of Brown-Séquard finds support.
>
> As further evidence of the effect produced upon the nervous system by the removal of the testes, it has been noticed in a number of cases that the patients afterwards suffer from uncomfortable flushes of heat, similar to those experienced by women at the time of menopause. Also distinctly hysterical phenomena have been observed after castration.[22]

Cabot's comments are also illustrative of the ways in which the medicalization of male sexuality was coextensive with the pathologization of the bodies of women. As women were increasingly assigned to the private, domestic, and emotional sphere of life based upon their biological nature and limitations, so men who could not exercise their reason or practise restraint risked descending from the masculine to the feminine state of being. In a fate worse than death a disruption of the natural sexual state of men risked their metamorphosis into the mental and physical state of womanhood; dire stuff indeed.

Beyond the debate over castration, urologists were also struggling with the thorny issues inherent in creating iatrogenic problems for the sexual economy of the male as a result of their surgical interventions.[23] Suprapubic and perineal approaches to prostate surgery were weighed against each other and the potential harms and benefits as were other kinds of interventions such as the 'galvano-surgery' pioneered by the Italian surgeon Enrico Bottini that found considerable, if short-lived, fame. Debates revolved around the best ways of maintaining the integrity of the ejaculatory ducts and the preservation of erectile function, but such considerations were not 'merely' a matter of the psychological impact of impotence but rather an attempt to protect the systemic health of the male. This was a period of great change in the very concept of marriage, at least for the middle classes.[24] The nineteenth century rise of the so-called 'companionate marriage' with its emphasis on romantic love (rather than, say, financial expediency) focused attention on the role of healthy sexuality within a healthy marriage; impotence in the male and frigidity in the female was to be assiduously avoided. Surgeons wishing to take a scalpel to the male sexual organs then did so with the certain knowledge that much rested upon their skills and choices of which techniques and devices to employ over others.

PROSTATE HEALTH AND TESTICULAR EXTRACT IN THE MEDICAL MARKETPLACE

This state of medicalized and sexualized masculinity was not just a matter for doctors. As Hitchcock had warned, media and other forms of popular discourse advertised both the 'problem' and its 'cure', helping to create a market as ripe for exploitation by the ambitious as it was highly attractive (if bewildering) for the anxious male consumer seeking to improve his life. Censorious attitudes surrounding sexuality and masturbation had turned discretion into a commodity. Prudishness on the part of the regular practitioner was harmful to the patient, Overall said, and threatened to create a vacuum that would be filled by the 'charlatan':

> The evil effect of masturbation upon the prostate and vesicles primarily, and the nervous system secondarily, has been over-estimated by many, and treated with too much indifference by others. The fact of the almost universal practice, at some time in life, among males, renders it a convenient source to which to attribute all the sexual and nervous diseases, not traceable to that of gonorrheal origin.
>
> Charlatans reap a rich harvest among youths and, too, older men, who being over-sensitive, are too prudish or secretive to consult their family phy-

sician and fall an easy victim to their tenets and ruse. The family physician, too, is often accountable for this, by not making a thorough examination of the case when consulted, treating the matter with too much indifference, dismissing him with a tonic, or telling him it is 'all in his head.' The fact is that most of those addicted to the habit are so ashamed of it, that they will deceive the physician, in the large majority of instances, by denying the practice altogether, or minimize the extent of indulgences so as to mislead him.[25]

As medical men and women continued to organize and professionalize, claiming both a moral and intellectual authority over who should and should not be properly considered a 'doctor', the pursuit of profit by 'irregular' practitioners was an unwelcome distraction. Much as Hippocrates had decried the itinerant, money-oriented 'sophists' of ancient Greece, so the new elites of medicine decried the 'quackery' of practitioners who did not adhere to the institutionally regulated behaviours of the late nineteenth century physician. How to distinguish the respectable from the disreputable was not obvious and it was not easy, but it was a problem that, as described above, Brown-Séquard attacked with true Bernardian flair. His persistent rationalization of the use of testicular extract as a modern 'scientific therapy' rested largely on his use of experimental animals to determine not only the efficacy but also the underlying 'dynamogenic' mechanisms of therapeutic action.[26]

By no means was everyone convinced by these experiments, and medical journals in the 1890s were replete with letters and articles objecting to the outlandish ideas and perceived exaggerated claims of Brown-Séquard.[27] Nevertheless, the pages of US and European medical journals of the 1900s and 1910s continued to discuss (and advertise!) 'testiculin', 'orichidin', and similar compounds prepared in accordance with the ideas of Brown-Séquard. Indeed, some practitioners recommended testiculin type therapies for everything from influenza to hysteria, diabetes to gangrene, precisely the kind of promotion of a 'miracle cure' that tended to ruffle the feathers of the AMA and other institutions of establishment medicine.[28] That said, it was not simply clinicians at the fringes of 'respectable' medicine that were turned on to Brown-Séquard's ideas. As the historian Chandak Sengoopta has argued, contemporary scepticism and derision aside, organotherapy stimulated a great deal of serious academic, experimental research into the glandular functions.[29] William Bayliss and Ernest Starling, for instance, colleagues from the Department of Physiology at University College London and famous for their work uncovering the role of the thyroid in health and disease, encouraged the use of thyroid extract as an all-purpose tonic to combat ageing and the stresses of life. Like other 'biologicals'—vaccines, serum therapies, and so on—these 'organothera-pies' seemed to offer a tangible, practical benefit of physiological research. Thanks to the high profile condemnation of physiology and physiologists by anti-vivisectionists in the late nineteenth and early twentieth centu-ries, these therapeutic 'successes' (several biologicals, especially the vaccine therapies, would eventually turn out to be duds) were all the more impor-tant to the status and reputation of the experimental science.

Hopes for an organotherapy panacea persisted during the interwar period but, more tangibly, the use of biological and the physiological modelling that underpinned this use lead to some other remarkable break-throughs. The discovery of the pancreatic extract in the 1920s that would eventually become to be known as insulin, for instance, was a model exam-ple of the use of physiology, vivisection, and experimental medicine in the creation of therapeutic advancement.[30] Of less historical importance but still notable was the continuation of work by experimentalists of Brown-Séquard's theories of the testes. In work that harked back to the studies of John Hunter and Louis Auguste Mercier on the effects of castration in the normal and diseased male, the Viennese physiologist Eugen Steinach and his surgeon colleague Robert Lichtenstern experimented with testicular implantation in guinea pigs. Thus in the 1910s Steinach and Lichtenstern continued to expand on the work of their nineteenth century predeces-sors and, via vivisection, proposed a series of specific ways that testicular secretions affected the development of secondary sexual characteristics in the male (in keeping with the moral attitudes of the times, they also pro-posed a therapeutic use for their implantation technique in this case for the 'treatment' of homosexuality in men). Sigmund Freud and the poet W.B. Yeats were two famous beneficiaries of the 'Steinach rejuvenation operation', in the mid-1920s and mid-1930s respectively. This procedure was not a grafting technique but rather a kind of vasectomy aimed at, so the theory went, reducing the burden of sperm production on the testes in order to boost the release of other testicular secretions.[31] Another kind of technique practised by the French surgeon Serge Voronoff involved graft-ing monkey testicles onto human donors, which, unsurprisingly enough, attracted the rapt attention of the lay and medical media alike.[32]

This period of experimentation was largely devoid of the kinds of con-trols that we now take for granted in experimental medicine—institutional oversight, informed consent, and so on—as the necessary basis of 'ethical' practice. For centuries physicians had generally decided for themselves (or with a group of peers) what could and could not be ethically done with their patients. So we have instances like the series of experiments performed by Leo Stanley, physician to the California State Prison, in which he along with his colleague, G. David Kelker, performed over a dozen testicular implants between animals and humans (and from executed to live prisoners) and more than six thousand subcutaneous abdominal injections of pulverized testicular matter in some four thousand prisoners in the years between 1918 and 1931.[33] As the historian Ethan Blue points out, Stanley's career at San Quentin, which lasted, as did the experiments, until 1951, was a curious coming together of several early twentieth century trends. Foremost amongst these was the great Progressive Era belief in the role of the expert as the pivotal factor in shaping a better future. This optimism partly explains how and why the next trend that underpinned Stanley's work—eugenics—was so widely and enthusiastically embraced as it was by public and professionals alike. For many, the notion that modern medicine could help rid mankind of physical and mental 'weaknesses', including the tendency toward fecklessness and criminality, was a heady proposition. For its advocates, the experiments at San Quentin (as with others of its kind) provided the dual purpose of allowing prisoners to 'repay their debt' while also providing a social benefit to the sick and, ultimately, a means though which we might improve upon the masculinity and virility of all 'healthy' men (eugenically speaking).[34]

Reaction to the atrocities of eugenic exercises and human experimentation in the name of Nazism, and the journalistic exposes of research on vulnerable, marginalized populations during the Cold War, would eventually lead to a new kind of ethics, a 'bioethics', but all that was in the future.[35] Suffice it so say that doctors in the early-to-mid twentieth century inhabited a quite different ethical frame than those in the late century period. As uncomfortable as might be to think about, such context is worth keeping in mind if we are to understand how 'ordinary' practitioners engaged in such extreme experiments. This was a time when clinicians routinely practised a kind of 'benevolent deception' with patients, motivated by a belief that not allowing the patient to know of a terrible diagnosis, like cancer, was a continuation of the therapeutic responsibility. By the 1960s and 1970s the legal, political, and cultural push back against such paternalistic forms of authority overwhelmed the rationale but did not change the fact that what was 'ethical' to a clinician in the 1920s might be incommensurate with what was ethical to a clinician fifty years later.[36]

Another issue in the 'ethical blurring' for clinicians like Stanley, though, was the fact that, as mentioned above, patients routinely sought out and paid for similar treatments from the kinds of practitioners that he, along with other 'respectable' establishment physicians thought of as 'quacks' and 'charlatans'. The medical breakthroughs of the late nineteenth century were big news, and popular magazines and newspapers eagerly reported on the latest tales of innovation and medical 'breakthroughs'.[37] The fashion for testiculin had added a dose of titillation to the entertainment, a feature that was not lost on the practitioner (whether licensed or not) wishing to make a buck. Throughout the interwar period, for instance, to the great frustration of the AMA, John R. Brinkley—also known as the 'goat-gland doctor'—possessed a thriving private practice in Kansas, promising men sexual rejuvenation and cure of prostate problems and prostate cancer via surgical grafting (of goat testicles) and other proprietary techniques. The AMA were appalled with Brinkley as much for his salesmanship as for his science, particularly as he used one of the first radio stations to broadcast state wide, Kansas' KFKB, to promote his cures and to provide advice on air to would-be patients.[38]

What of the patient in all of this, however? How did the advances in prostate treatment influence his view of sex and sexuality? Medical advice manuals of the type authored by Howe and Overall in the 1880s and 1900s respectively had, by the 1930s, begun to consciously seek out lay audiences. It should be noted here that medical treatises aimed at a popular audience were nothing particularly new. As the historian Helen Yallop shows, popular works with broadly medical themes were available from the sixteenth century onwards.[39] While these early texts were usually poorly written, cheaply produced, efforts, the genre of medical advice literature blossomed in the eighteenth century. This was, Yallop argues, partly because medical elites wished to popularize and domesticate their visions of the body and health at the expense of other, informal, healing traditions, and partly because the educated classes increasingly sought to gain control over their lives and fates through the acquisition of knowledge in that Age of Enlightenment.[40] As the nineteenth century progressed, though, manuals that dealt with sexual and marital hygiene had to contend with accusations of promoting quackery as the line between populist medicine and sage advice became increasingly blurred.[41] The ongoing pathologization of masculinity and male sexuality further complicated the issue as a man likely became less, not more, inclined to seek the advice— and possibly face the opprobrium—of his regular physician.

A 1935 medical advice manual, *The Dangerous Age in Men: A Treatise on the Prostate Gland*, written by the New York Bellevue Hospital urologist Chester Stone picks up these themes with an opening exposition of the importance of an informed layman:

> Despite the modern trend away from secretiveness and ignorance there still are many facts pertaining to man's physical and mental well-being that have been neglected. The prostate gland remains an unexplored country to the average layman. It is this gland which causes man's mental and physical suffering during his dangerous period.
>
> It is with this purpose of describing this all important gland, its changes and functions, of developing an honest knowledge of the rules necessary to preserve it from irritation and disease, or, if these have occurred, the precautions then to be observed, that this monograph is written.[42]

Premised on the notion that, '[n]o man is … old if his sex glands are active',[43] Stone's book sought to guide men to a well-ordered sexual life in the face of age and disease. While the causes of prostatic 'congestion' differed little between Stone's account and those of Howe and Overall—masturbation, too little sex, too much sex, overly prolonged sex, and so on—the optimism for a return to a state of health, if properly treated by a competent urologist, are strikingly different in the later account. For Stone, few causes of prostatitis (perhaps even cancer) were hopeless so long as the patient engaged his doctor at the earliest signs of mental or sexual dysfunction, and so prevented the progression of disease.[44] Echoing the words of Hitchcock a century earlier, Stone argued that shame and fear (of surgery or of a diagnosis of venereal disease), fundamentally threatened the health of men. A later book published in 1950 by another Bellevue urologist Herbert Kenyon, similarly decried how ignorance and embarrassment were killers and, in a similarly perennial fashion, reviled the vulnerability of men to the predations of charlatans:

> In many circles open discussion of conditions of the genital and urinary organs is socially unacceptable, and men troubled by symptoms of prostatic derangement hesitate to seek advice from friends and even from members of their own family for reasons of delicacy. The field is therefore wide open to purveyors of misinformation and those who deliberately exploit the public's ignorance of the prostatic diseases.
>
> There is a profusion of commercial advertising in the cheaper magazines and some newspapers throughout the country, urging the purchase of quack remedies and devices for relief of the symptoms of "Prostate Trouble."[45]

While Kenyon's account is devoid of references to the sexual neurasthenia that so occupied his predecessors, the supposed mechanical damage caused by masturbation and other sex acts outside of 'routine' heterosexual intercourse are held to be suspect. Informed by the new science of prostate cancer represented by the work of Charles Huggins (described below), Kenyon dismissed any suggestion of a causal relationship between sexual dysfunction and the development of prostate cancer.[46] The publication of Alfred Kinsey's *Sexual Behavior in the Human Male* two years earlier in 1948 had made the case for the normalization of masturbation. Although he was by no means the first to make the case for this, the immediate scandal and notoriety that *Sexual Behavior* attracted (in large part due to its introduction of the 'Kinsey scale' which placed homosexuality on a continuum with heterosexuality, and not, as many others had argued as a discrete pathological category) helped to circulate these ideas more widely. As Kenyon's account shows, however, while the moral panic over masturbation might have subsided somewhat by mid-century, the notion that the prostate could be susceptible to excessive strain persisted.

Kenyon's account is interesting in another way too. While he was realistic about the limitations of urologists in the face of some prostate diseases, particularly cancer, Kenyon's principal concern was to urge *all* men to place themselves under the care of a urological specialist as a preventative measure against future problems. 'Health', Kenyon argued, 'does not just happen, but must be actively sought.'[47] If we are to take Kenyon's views as representative, urology had certainly come a long way in the forty years since Cabot had expressed his cautious optimism about the possibility that urology might be considered its own specialty. Thanks to a new focus on prostate cancer, the prominence of the urologist was set to rise much further.

By the 1930s the study of biological secretions, or 'hormones' as Starling (at the suggestion of his physiologist colleague, William Hardy[48]) had designated them drawing from the Greek word *hormôn*—meaning 'to set in motion' or 'impetus'[49]—were all the rage. The 'rage' was, however, a paradoxical admixture of the most highbrow laboratory work, with the ethically questionable, with the most derided of commercial hucksterisms. It was quite a legacy, but as American medicine became more regularized by mid-century so research into sex hormones lost the sensationalism and the tinges of quackery. This was particularly the case from the 1940s onwards, when new synthetic 'androgens' began to replace testiculin and other organic preparations as the basis of hormone treatment thanks, in large part, to the identification and isolation of one very potent ingredient: testosterone.

Sixty years on from the publication of Bernard's *An Introduction to the Study of Experimental Medicine*, the success of insulin helped to show the world what was possible when physiological research, clinical expertise, and the manufacturing resources of the pharmaceutical industry came into alignment. More immediately for scientist and industry, it showed how biologicals could offer grand opportunities for international academic prestige and serious commercial profits. The area of hormones in particular became a hive of scholarly activity. The German scientist Aor Windaus received the 1928 Nobel Prize in chemistry for his hormone and vitamin work, for instance, and a year later his student Adolf Butenandt isolated the first 'sex hormone', estrone, from the urine of pregnant women (a feat also completed independently by the St. Louis chemist, Edward Doisy). A few years later a substance Butenandt labelled 'androsterone' was similarly isolated from the urine of hundreds of policemen.[50] Other researchers soon discovered that androsterone was not responsible for the production of secondary sexual characteristics in males but this early work of Butenandt provided a glimpse into what was possible in hormone studies.

Numerous high profile US and European laboratories worked on isolating sex hormones in the 1920s and 1930s. It was the Austrian physician Ernst Laqueur, however, who, using his connections with industry, helped to create a new paradigm in sex hormone research in the interwar period.[51] Laqueur first developed a partnership with the Dutch company Organon (itself founded in 1923 on the hope that animal extracts could be manufactured as medicines) in order to produce insulin. With the success of that operation Laqueur's group at the University of Amsterdam followed similar procedures of hormone extraction, only this time using bovine testicles rather than porcine pancreases. In 1935 the team published the now classic paper, On Crystalline Male Hormone from Testicles (Testosterone),[52] choosing the name 'testosterone' to signify the relationship of the hormone to the testes. The word 'testosterone' also made use of a convention established by of Butenandt to name hormones with 'ster' for sterol, and 'one' for ketone.[53] Realizing that his androsterone was actually a metabolite of the new substance, a year after Laqueur's publication Butenandt worked out the chemical composition of testosterone. This and similar work opened the way for Organon and other manufacturers to begin producing large quantities of synthetic testosterone by the late 1930s. For his achievements in hormone research Butenandt was awarded the 1939 Nobel Prize in Chemistry (Doisy was not recognized for his role in the isolation of hormones, although he did go on to win his own Nobel Prize—in Physiology or Medicine—in 1943 for his role in isolating Vitamin K).

The decades following Brown-Séquard's death were then an extraordinary golden age of 'organotherapy'. The best minds and the best laboratories in the world fiercely competed for primacy in research, and following on the successors of Organon, drug houses turned over money and other resources in the race to commodify biological products.[54] For one young Canadian researcher, Charles Huggins, the study of glandular secretions would lead to ground-breaking work on the relationship between cancer and hormones for which he too would be awarded a Nobel Prize (in 1966 for Physiology or Medicine). For Huggins and a generation of physiologists and (the newly coined designation of) 'biochemists' the industrial production of hormones was critical to their research programs as was the increasing number of academic, medically orientated laboratories that sustained their careers.

CHARLES HUGGINS AND THE TREATMENT OF PROSTATE CANCER WITH HORMONES

When Huggins arrived at the University of Chicago in the 1920s the school of medicine was a very new enterprise, both literally and figuratively. Like Johns Hopkins before it, Chicago adopted a model of salaried research-physicians that put its academic clinicians outside of the established referral networks in the city. As Huggins himself remembers it, it also created the fear amongst the public that if they did go to the hospital they might be experimented on by the research scientists.[55] As Huggins later explained, this dearth of patients had him, as a junior faculty member in urological surgery (a subject in which he had little practical experience), casting about for a research problem to work on and he finally settled on working with hormones and the prostate.[56] During this time of early departmental development Huggins' chief of surgery, Dallas B. Phemister, sent him to London to study with the Nobel laureate chemist, Robert Robinson, to learn the arts of the new biologically informed chemistry, or 'biochemistry' as it was coming to be known. It was here in Robinson's laboratories that Huggins learned a love for chemical markers and 'objective', quantifiable, indicators that would become so crucial to his own way of thinking.

In part Huggins chose the prostate as the subject of his researches because of the availability of an experimental animal model, saying of this in his typical tongue-in-cheek fashion, 'There is a high incidence of abnormal growth processes—of tumors in the prostate gland of certain species in senescence. These species are man, the dog and the lion. For technical reasons, observations can be carried out with greater facility in the first two types.'[57] For an ambitious young surgeon, though, the prostate gland had other attractions. By the early 1930s carcinoma of the prostate was increasingly recognized to be not as rare as turn of the century clinicians had assumed, but rather a common condition especially in older men. What had not changed though was the belief that the condition was very hard indeed to treat surgically.[58] Prostatectomies were still performed, although more for palliation than as a meaningful attempt at cure, and, as I will discuss in Chap. 7, some initial attempts at irradiating the prostate had rendered mixed to poor results. In 1926 a study of one thousand cases seen at the Mayo Clinic concluded that regardless of treatment two-thirds of patients diagnosed with prostate cancer died within nine months.[59] There was, in other words, a certain ongoing despondency surrounding prostate cancer, and relatively little for an experimentalist like Huggins to work on surgically. As an experimental *physiologist*, however, Huggins could and did tap into the still roaring intellectual movement of hormones in the hope of discovering more about their role in the normal and abnormal function of the prostate.

In a paper published in 1939, Huggins reported a slew of quantitative data on the prostatic secretions of dogs that had undergone a surgical procedure to isolate the prostate.[60] The researchers were able to stimulate the gland via administration of pilocarpine hydrochloride (a substance active in stimulating the parasympathetic nervous system) that then resulted in secretions collected in flask attached to a cannula connected to the prostate of the dog. The experimental surgery was also conducted on castrated dogs, and dogs that had had their thyroid and parathyroid glands removed, and the chemical composition of the samples taken from the two groups of dogs were then compared side by side. Huggins showed that the prostate gland in the castrated dogs could be restored to near normal function using testosterone injections and so was able to disprove a theory that the thyroid played a major role there. Although his findings were rather modest in that respect, the paper set out in some considerable detail the experimental preparation of an animal model, as well as methods of chemical analysis of prostatic secretion, that, when combined with histological analysis of prostate tissue, would lead to Huggins' great breakthrough.

By the mid-1930s it was well described in the scientific and medical literature that prostate function depended in some significant way upon secretions from the testes, which in turn depended upon a range of other endocrine products for normal function. By this point, researchers had also isolated several different hormones that were linked to the production of secondary sexual characteristics, which they had labelled 'androgens' and 'oestrogens'. Furthermore, a consensus had emerged that androgens (including testosterone) and oestrogens were together responsible for the normal functioning of the prostate and that abnormalities could arise from imbalance between the two hormones.[61] Huggins knew that androgens in excess caused hyperplasia (enlargement) of the cells of the epithelium of the prostate and an increase in prostatic secretions, while oestrogens caused a metaplasia (cellular alteration) and so a decrease in the function of the secretory epithelia of the prostate. In other words, the action of androgens on prostatic size and secretion seemed be antagonized or negated by the action of oestrogen but nobody knew why this was the case. In 1940 Huggins published a continuation of his prostate studies using naturally occurring benign prostatic enlargement in senile dogs as a modification of his original experimental model.[62] Once again, Huggins castrated the dogs but this time he treated them with injections of androgenic *and* estrogenic matter. Using his prostatic isolation method, Huggins and his collaborator Philip Clark, collected and analysed the prostatic secretions over several months, building up a detailed chemical picture of the functional, physiological, life of the hypertrophied prostate. Also in 1940, the prolific Huggins published another paper essentially repeating the experiments conducted by William White, Arthur Cabot, and others at the end of the previous century, aimed at uncovering the connection between castration and benign hypertrophic prostate disease.[63] Unlike White and Cabot, Huggins reserved his surgery for the laboratory's dogs, but his own crossover to human studies was by then well under way.

A year later Huggins and his student, Clarence Hodges, published what would become a seminal paper in the literature of twentieth century scientific medicine, Studies on Prostatic Cancer: I. The Effect of Castration, of Estrogen and of Androgen Injection on Serum Phosphatases in Metastatic Carcinoma of the Prostate.[64] The key to Huggin's breakthrough was, he believed, reliant on his quantitative, chemical approach to problems of disease in a research world still wedded to descriptive pathological anatomy. As he said in the later reflection on the impact of his work published in 1943 (in words strikingly reminiscent of Claude Bernard):

The functional approach contrasts sharply with the descriptive approach—with the methods of classical pathology. ... Assay of a disease in a laboratory obviously removes much of the uncertainty inevitably associated with bedside observation, particularly in cancer. The yardstick in prostatic cancer concerns certain enzymes, the phosphatases.[65]

Thanks to his physiological studies with dogs and his extensive breadth of knowledge of the biochemical literature, Huggins felt certain that phosphatases could be used as a reliable marker freeing him from a reliance of the 'subjective' and 'descriptive' indicators of disease at the bedside of the human subject. In later writings, Huggins made casual mention of stumbling on this area of prostate cancer research somewhat by accident. He reported that during his early metabolic studies he had first seen elderly dogs with prostatic cancer as an interference with his work. He would soon see such animals as a great opportunity to connect his research to the wards.

Whatever the origin story of his interest in cancer, it seems highly likely that the nearby patients of a new medical school, unencumbered as they were with the entrenched traditions and hierarchies of more established institutions, were a huge draw for Huggins. Certainly it was a step that other research clinicians inspired by his work were making,[66] and it seems likely that a man of Huggins' ability and ambition was keen to push his physiological research methods into clinical practice. The modern setup of the Chicago school with its availability of both laboratories and inpatients allowed for a dazzling series of experiments in which Huggins and his co-workers castrated and then followed patients with advanced prostatic cancer (metastasized to the bone). The results of the experiment showed that a reduction in androgens—whether from surgical castration or so-called 'chemical castration' achieved by the anti-androgen effects of diethylstilbestrol or DES (an oestrogen compound) —caused a reduction in activity of carcinomas of the prostate as indicated by reduced levels of phosphatase. In addition to a measurable biochemical effect, Huggins in the second paper of the prostatic cancer series, 'Studies on Prostatic Cancer II: The Effects of Castration on Advanced Carcinoma of the Prostate Gland',[67] also published in 1941, noted the improved clinical picture of men treated via a reduction of androgenic activity, including reduction in pain and an increased energy and appetite. In another illustration of the different ethical times in which Huggins worked, Chicago patients were subjected to repeated biopsies and x-rays to record details about the behaviour of the neoplasm and any bony metastases. Huggins and his team also tried injecting androgens back into patients who were undergoing relief to see if the pain came back; it did. Such intrusions are perhaps better understood if we again consider Huggins' attitude towards 'objective' indicators of biological activity. His writings frequently show discomfort about any perceived 'subjective' measures. It is significant then, I think, that he ensured that bedside observations were 'corroborated' with changes in precisely measured weight and examination of red blood cells for decreases in the anaemia that often accompanied neoplastic disease. For Huggins, quantification was all.

In spite of his successes, Huggins was entirely candid about the fact that castration—whether chemical or surgical—was not a *cure* for prostate cancer, but he did encourage optimism about the easing of symptoms and even some temporary tumour remission for the majority of men treated, saying: 'it is possible by reducing the amount or the activity of circulating androgens to control, more or less but often extensively, far advanced prostatic cancer in large numbers of patients'.[68] The reason why Huggins won the Nobel Prize for this series of experiments was that he helped overturn the notion of the 'autonomous' cancer cell, a view that regarded malignancies as self-perpetuating with little or no extraneous influence. Huggins showed that for some cancers at least, neoplastic growth was hugely dependent on circulating hormones, on the *milieu interior* in other words. As such, Huggins' work opened up new possibilities for the treatment of cancer, especially cancer that was considered too advanced for surgery. As Huggins himself said in his Nobel acceptance speech, the study of prostate cancer was the start of chemically targeted therapies in *all* kinds of malignant disease.[69]

In the years that followed the publication of Huggins' original work on the prostate, clinicians across the US enthusiastically tried it on their own patients. As two such academic clinicians from the University of Michigan, Reed Nesbit and William Baum, noted in the pages of the *Journal of the American Medicine Association* 1950, though, the problems with the method were soon apparent:

The physician has seen the original enthusiasm attendant on this discovery become tempered by the knowledge that the benefits were not universal, that the degree and duration of response were variable and that eventually most patients experienced relapse and subsequent death from the primary disease. The physician has been confronted with problems concerning the selection of the form of endocrine modification most efficacious for the particular needs of the patient, the designation of the most opportune time to institute therapy and the choice of secondary therapy once relapse has occurred.[70]

For Nesbit and Baum, the most logical way to answer these questions was to try to draw together the medical records of the thousands of patients who had undergone endocrine treatment in the decade since the publication of Huggins' article. Using a grant from the United States Public Health Service (USPHS), Nesbit and Baum devised a standardized data form for use by participating urologists to record diagnosis, therapy, and follow-up of their patients. Their retrospective study of the one thousand eight hundred and eighteen records obtained included patients treated with DES, or with surgical castration, or both. Patient records were then further divided between cases showing the presence or otherwise of metastases—itself a determination made by reviewing records for phosphatase levels, evidence of bony masses in x-ray images, and biopsy results showing the extent of lymph node involvement.[71] Nesbit and Baum retrospectively compared the three- and five-year survival rates of the different groups with an historical control group of untreated patients. The overall improvement in survival times they observed led them to come out strongly in favour of oestrogen control in the treatment of prostate cancer. In their view oestrogen treatment (utilizing both DES and orchiectomy) would be optimally employed at first diagnosis so as to avoid as long as possible the production of metastases, which they associated with an extremely poor prognosis.[72] It was known to Nesbit and Baum, as it was known to other researchers, that over time prostate tumours became 'androgen independent' meaning that endocrine therapy only worked for a certain period of time, but they recognized that this timeframe could often differ quite widely between ostensibly similar looking cases and so saw no reason for excessive pessimism about the treatment approach.

As a mid-1950s letter titled to the editor of the *Journal of the Association of American Medicine* makes clear, though, vestiges of the past continued to be observed in this 'new age'. Hormones were still regularly added to vitamin pills and recommended for postmenopausal women and men over fifty, regardless of the by then known links between hormones and prostate cancer in the latter. 'Four times in this past week', the writers—a group of Pennsylvanian clinicians—complained:

we have received advertising folders from pharmaceutical firms highly recommending that we ... prescribe our patients [testosterone] ... for rather
vague conditions in older men, and recently there has been a wave of enthusiasm for a combination of androgenic and estrogenic substances presumed
to provide a 'balanced' hormonal therapy especially beneficial in geriatric
practice. ...

Despite the happy prospect of bolstering up a few sagging metabolic
processes or doing a little diffuse toning up of tissues here and there for
our patients over 50, we are disinclined to reach for the prescription pad in
response to these rather sweeping claims.[73]

It was perhaps in response to the old commercialism in the new science
of disease that that the editor gave the letter the title, 'Of fires and frying
pans', but punning aside, the old hopes for panaceas and age old efforts to
exploit such hopes for commercial gain would feature heavily in the post-
WWII realities of American biomedicine.

The experimental model of medicine promoted by Claude Bernard and
other academic physiologists of the late nineteenth and early twentieth
century would go on to profoundly shape the growth of clinical research,
institutionally and intellectually. The desire to test the new drugs and iso-
lates developed in laboratories on patients in the wards was clearly most
feasible in scenarios where clinicians and bench scientists worked in close
quarters—as they did within a university setting. The complex needs of
clinician-researchers for laboratories, animals, and patients slowly established academic medicine as its own kind of practice, one whose leaders
would go on to become the elite of the medical and surgical profession
in the years following WWII. The high profile successes of the 'biologicals' during the interwar years doubtless helped this transition, as did the
considerable attention that pharmaceutical companies paid to developing
links to academic institutions.[74]

The career of Charles Huggins is a good example of this shift. He
began his career in a brand new if underused academic medical centre,
and ended it with successes joining the bedside, the laboratory, and pharmaceuticals. To his animal model for prostate cancer, he was later able to
add distinguished work on a rat model for breast cancer. As in the case
of prostatectomy and castration for prostate diseases, surgeons of the late
nineteenth and early twentieth century had attempted to treat diseases of
the breast via a removal of the ovaries, with similarly limited successes. So
it was that Huggins' rat model allowed for detailed biochemical work on

the role of hormones to be applied to the treatment of breast cancer and the use of antagonists to deter the growth and spread of tumours. It was also a model that would later be used by a team of British scientists at the Imperial Chemical Industries (ICI) in the early 1960s in the production of an oestrogen antagonist that would become known as Tamoxifen, one of the most significant anti-cancer drugs of the late twentieth century.

The world of medicine was changing in other ways too. In the conclusion of their study on treatment outcomes for prostate cancer, Nesbit and Baum made the following observation in reference to their research methodology:

> This study indicates the value of cooperative effort in the acquisition of data of statistical significance, where previously the handicaps of time and numbers had made the accumulation of such information an individual impossibility. This economy might well be applied to other fields of investigation in which, in the interest of accurate evaluation, the study of large numbers of patients is a necessity.[75]

This 'economy' as they called it, was indeed on the minds of many. WWII-era successes in the treatment of infection and infectious diseases had helped inspire another line of inquiry into cytotoxic drugs, or chemotherapies, inquiries that were soon turned to the study of cancer. The study of these new therapies (with or without surgical, immunological, and hormone approaches) would go on to drastically alter the ways in which the clinical applications of biomedical research were investigated. The era of the randomized clinical trial was about to begin.

NOTES

1. Overall, *A Non-Surgical Treatise on Diseases of the Prostate Gland and Adnexa*, 151.
2. Huggins, Endocrine Control of Prostatic Cancer, 541.
3. Bernard, *An Introduction to the Study of Experimental Medicine*, 2.
4. Brown-Séquard, Note on the Effect Produced on Man by Subcutaneous Injections of a Liquid Obtained from the Testicles of Animals.
5. Freeman, Bloom, and McGuire, A Brief History of Testosterone, 371.
6. Brown-Séquard, Note on the Effect Produced on Man by Subcutaneous Injections of a Liquid Obtained from the Testicles of Animals, 105.
7. O'Shea, A Plea for the Prostate, 11.

8. Overall, *A Non-Surgical Treatise on Diseases of the Prostate Gland and Adnexa.*
9. Ibid., 34.
10. For a detailed examination of the concept of 'neurasthenia' in different cultures and countries see Gijswijt-Hofstra and Porter, *Cultures of Neurasthenia from Beard to the First World War.*
11. A comprehensive account of the historical meanings of 'hysteria' can be found in Showalter, *Hystories.*
12. Overall, *A Non-Surgical Treatise on Diseases of the Prostate Gland and Adnexa*, 153.
13. Hodges, History of Sexual Medicine, 725–6.
14. Hitchcock, Insanity and Death from Masturbation, 286.
15. Mumford, 'Lost Manhood' Found, 45–6.
16. Ibid., 47.
17. Stephens, Pathologizing Leaky Male Bodies, 423.
18. Rosenman, Body Doubles, 365.
19. Howe, *Excessive Venery, Masturbation and Continence*, 67–8.
20. Ibid., 68–9.
21. Ibid., 252.
22. Cabot, II. The Question of Castration for Enlarged Prostate, 198.
23. O'Shea, A Plea for the Prostate, 22–3.
24. For a discussion of middle class ideas about marriage in early twentieth century American see Davis, 'Not Marriage at All, but Simple Harlotry'.
25. Overall, *A Non-Surgical Treatise on Diseases of the Prostate Gland and Adnexa*, 35.
26. Brown-Séquard, Note on the Effect Produced on Man by Subcutaneous Injections of a Liquid Obtained from the Testicles of Animals.
27. Aminoff, *Brown-Séquard an Improbable Genius Who Transformed Medicine*, 240–2.
28. Freeman, Bloom, and McGuire, A Brief History of Testosterone, 371.
29. Sengoopta, 'Dr Steinach Coming to Make Old Young!', 122.
30. Porter, *The Greatest Benefit to Mankind*, 566–8.
31. Schultheiss, Denil, and Jonas, Rejuvenation in the Early 20th Century.
32. Hamilton, *The Monkey Gland Affair.*
33. Stanley, Testicular Substance Implantation.
34. Blue, The Strange Career of Leo Stanley.
35. For a superb account of the evolution of the idea of medical ethics in the US see Baker, *Before Bioethics.*
36. This is not to say that there was no variation in views on what constituted ethical practice. The work of Susan Lederer provides some great insight into these kind of debates: see, for instance, *Subjected to Science*, 13–15.
37. Hansen, New Images of a New Medicine.

38. Lee, *The Bizarre Careers of John R. Brinkley.*
39. Yallop, *Age and Identity in Eighteenth-Century England*, 13.
40. Ibid., 11–12.
41. Stephens, Pathologizing Leaky Male Bodies, 435–6.
42. Stone, *The Dangerous Age in Men: A Treatise on the Prostate Gland*, v.
43. Ibid., 4.
44. Ibid., 65.
45. Kenyon, *The Prostate Gland*, viii–ix.
46. Ibid., 141.
47. Ibid., 172.
48. Freeman, Bloom, and McGuire, A Brief History of Testosterone, 372.
49. Starling, On the Chemical Correlation of the Functions of the Body.
50. Freeman, Bloom, and McGuire, A Brief History of Testosterone, 372.
51. For an overview of this period see Oudshoorn, *Beyond the Natural Body.*
52. David, Über Krystallinisches Männliches Hormon Aus Hoden (Testosteron).
53. Freeman, Bloom, and McGuire, A Brief History of Testosterone, 372.
54. Gaudillière, Better Prepared than Synthesized.
55. Walsh, How Charles Huggins Made His Nobel Prize Winning Discovery.
56. Ibid.
57. Huggins, Endocrine Control of Prostatic Cancer, 541.
58. Barringer, Carcinoma of the Prostate.
59. Ibid.
60. Huggins et al., Quantitative Studies of Studies of Prostatic Secretion.
61. Editorial, Endocrinology of the Prostate.
62. Huggins et al., Quantitative Studies of Studies of Prostatic Secretion.
63. Huggins and Stevens, The Effect of Castration on Benign Hypertrophy of the Prostate in Man.
64. Huggins et al., Studies on Prostatic Cancer.
65. Huggins, Endocrine Control of Prostatic Cancer, 541.
66. Herbst, Biochemical Therapeusis in Carcinoma of the Prostate Gland.
67. Huggins et al., Studies on Prostatic Cancer.
68. Huggins, Endocrine Control of Prostatic Cancer, 544.
69. Huggins, Nobel Lecture, 240.
70. Nesbit and Baum, Endocrine Control of Prostatic Carcinoma; Clinical and Statistical Survey of 1818 Cases, 1317.
71. Ibid., 1318.
72. Ibid., 1320.
73. Roth, Hess, and Kaminsky, Of Fires and Frying Pans.
74. See, for instance, Rasmussen, The Moral Economy of the Drug Company-Medical Scientist Collaboration in Interwar America.
75. Nesbit and Baum, Endocrine Control of Prostatic Carcinoma; Clinical and Statistical Survey of 1818 Cases, 1320.

CHAPTER 5

Cancer and Clinical Trials

A major problem in randomized clinical trials of anticancer agents lies in the great variation in the natural history of cancer, necessitating the use of many patients in different test groups to average out this variation. If it were possible to predict the course of the disease and divide the patients into groups known to have a longer or shorter course of disease, the reduction within-group variation might help bring out differences in response to various procedures or agents.
Donald Gleason, Classification of Prostatic Carcinomas (1966)[1]

In a way then, the Cancer Program can be regarded as an unusual and fragile biological organism with a head at each end. One end is concerned with basic research and the other with the application of the results of basic research. The organism is unable to survive if either head is severed. Vincent DeVita, The Governance of Science at the National Cancer Institute (1983)[2]

Experiments, as Claude Bernard had shown, manipulate materials and processes in the natural world to determine underlying cause-and-effect relationships. To do this, experimenters seek to change certain known conditions and control for others, an inductive and powerful approach to problem solving. This drive for experimentalism, combined with the search for new therapeutic agents arising in the years before WWII, helped stimulate by mid-century perhaps the key methodological (we might even say ideological) innovation in twentieth century biomedicine, the randomized clinical trial (RCT). Although there are other claimants to the title of the 'first' controlled clinical trial, historians generally recognize the MRC's 1946 study of tuberculosis treated with streptomycin[3] as the beginning of

© The Editor(s) (if applicable) and The Author(s) 2016
H. Valier, *A History of Prostate Cancer*,
DOI 10.1057/978-1-137-56595-2_5

the new movement.[4] Part of the notoriety attached to the streptomycin trial came because its results were, unlike some earlier efforts, positive and clearly so.[5] The original scarcity of the antibiotic had made the division of patients between a treatment group who got the drug and other groups who did not ethically straightforward (a feature of clinical trials that would become increasingly complex as I discuss in the next chapter), and helped to make the results clear and compelling.

Bernard's desire for control and manipulation lies at the heart of this original vision of the RCT, with carefully selected groups of 'like' patients randomly placed into either 'control' groups, which might receive a biologically inactive 'placebo' or standard of care (or a mixture of the two), and 'treatment' groups receiving the intervention to be studied. Conscious and unconscious bias in placing patients, such as, for example, a tendency to place sicker patients into a known treatment group is thereby averted and statistical inferences can be drawn from clinical outcomes with a higher degree of certainty. Similarly, the variability of disease and the idiosyncrasies of its complications within the host that might militate against comparisons of therapeutic regimens were substantially counteracted because all patients received their treatment assignment by chance. Thus the casuistic and observational basis of western clinical medicine (established in the Hippocratic tradition) was fundamentally reconsidered within this new way of knowing and biomedicine would never be the same again.

Historians have argued over the extent to which RCTs became the 'gold standard' of research in biomedicine because they also in part supported a certain administrative way of thinking favoured in the post-WWII period.[6] The expansion of the National Institutes of Health (NIH) in the United States would be a good example of what I'm talking about here. The National Cancer Institute (NCI) was created with the passage of the National Cancer Institute (NCI) Act in 1937, a continuation of a commitment to ease social problems with federal money arising from New Deal politics. How cancer research should be conducted in practice was not, however, a settled issue. As the historian Gerald Kutcher notes, when the physician-researcher Gordon Zubrod left Johns Hopkins for the NCI of the NIH he did so at an auspicious time:

> Until the late 1940s, the old medical guard who controlled the National Advisory Cancer Council, a body that guided the NCI, strongly resisted the type of large-scale engineered research that had become popular during

the war [such as the antimalarial researchers that Zubrod had himself participated in during his service in the US army medical corps]. But, in 1953, prominent physicians and others, among them Sidney Farber of Harvard, Cornelius Rhodes of Sloan Kettering, and the philanthropist Mary Lasker, successfully lobbied Congress for a $1 million grant to the NCI to investigate the possibility of an engineered program to cure leukemia.[7]

Wartime mass production of penicillin had done a great deal to consolidate support around the 'engineered' model of research during the late 1940s, and opposition to this approach was short lived, at least at the NCI. The horrors of weaponized gas during WWI, for instance, led the US military to research possible antidotes for poisoned troops. During WWII, three Yale University scientists employed in this research, Milton Winternitz, Louis Goodman, and Alfred Gilman, discovered that nitrogen mustard had a surprising ability to knock out white blood cells. From this observation they speculated that nitrogen mustard might be used to treat 'liquid tumours' like leukaemia and lymphomas. Their theories were confirmed shortly thereafter when a lymphoma patient under the care of their clinical colleague, Gustaf Lindskog, achieved a marked improvement of symptoms following administration of the compound.[8] Screening compounds for all kinds of biological activity slowly became—then stayed— big business in twentieth century biomedicine.

The early history of the Sloan-Kettering Institute located in New York City is a good illustration of the changing nature of biomedical research in North America. The institute was founded in 1945 and built adjacent to the nineteenth century Memorial Hospital for Cancer and Allied Diseases, so bringing laboratories and besides into close connection. When the President of General Motors and a trustee of the hospital, Alfred Sloan, pledged the sum of four million dollars to set up the institute his director of choice was Cornelius Rhoads, a former WWII head of the Medical Division of the Chemical Warfare Service. While the choice was an obvious one given Rhodes' experience of handling huge budgets and complex organized research teams, the historian Robert Bud points out that, like Sloan, Rhodes also shared the belief of other major industrialist-benefactors of the age that the industrial model would be good for medicine. This is something Sloan underscored when he insisted that the new Institute should bear not only his name but also that of the Director of Research at General Motors, Charles Kettering.[9] Both Sloan and Kettering went on to play active roles in the running of the institute and the planning of its research programmes, as did a number of other prominent east coast industrialists and scientific administrators.

This is not to say that investigators working outside of this industrial model were no longer active or engaged with high profile research. Sidney Farber at the Children's Hospital Boston, for instance, was amongst the first to try out chemical compounds for the treatment of childhood cancer. Encouraged by research going on into the role of B vitamins in pernicious anaemia, he began to study how metabolic agents might be used against leukaemia.[10] To his dismay he found that administering the B vitamin folic acid actually accelerated the progress of the disease. Undeterred, he contacted Lederle Laboratories of Pearl River, New York with a request that they synthesize a compound that could be drawn into the cancer cells and block the action of the vitamin. His subsequent publication, sponsored by the NCI and appearing in the *New England Journal of Medicine* in 1948, was a landmark study as it, for the first time, showed that remission (albeit temporary) could be achieved in the cases of children dying of leukaemia.[11] A period of great enthusiasm for the 'chemical cure' (or chemotherapy[12]) for cancer was about to begin, but individual-inspired efforts, like that of Farber, undertaken on a small series of patient cases was about to give way to a more centralized and organized systems of trials involving first hundreds then thousands of patients.

The million dollars that Congress awarded to the NCI thanks to the lobbying efforts of Lasker and others had largely gone to setting up the new Cancer Chemotherapy National Service Center (CCNSC) in 1955. The CCNSC was part of a great expansion in drug screening on the part of the NCI, acting, as Zubrod put it, as 'a pharmaceutical house run by NCI but with the operation widely dispersed in industry and universities'.[13] The CCNSC did have another major role, however, and this was to bring into being a formal network of NCI clinical trials cooperative group programmes. Once again, intimate links between the laboratory and the wards were essential in these efforts. Work at the NCI on the leukaemia mouse model by Lloyd Law had shown that a combination of drugs given at the same time worked better at inducing cancer remission than did giving the same drugs sequentially. This so-called 'combination chemotherapy' approach (first demonstrated effective in tuberculosis trials of PAS and isoniazid) was taken up by Law's clinical colleague, James Holland, who devised a protocol for mercaptopurine and methotrexate use in children with leukaemia in 1953 at the NCI's newly opened Clinical Center.

Lacking any background in cancer research, Zubrod used his experience of chemotherapy trials in infectious disease (tuberculosis and pneumonia) and his knowledge of trials of analgesic drugs as a basis for his planning of future guidelines for cancer trials.[14] He reached out to James Holland, who had moved on to Roswell Park in 1954 before his trials at the Clinical Center were complete, and recruited him to work with a new face at the NCI, the physician-researcher Emil Frei. Between them, the three men organized what would be one of the first collaborative oncology trials in the US, drawing adult and paediatric patients from the NCI's Clinical Center, Roswell Park, and the Children's Hospital of Buffalo, New York.[15] The results of the trial showed an encouraging (if still temporary) remission of cancer in children with acute lymphocytic leukaemia and adults with acute myelocytic leukaemia. More than this, though, the researchers felt that their collaborative model was a powerful and widely applicable investigative instrument but only if undertaken with great care and foresight:

> The mechanics of a cooperative study conducted in different places by several individuals assumed critical importance. A printed protocol was distributed which specified procedural steps in detail. The conclusions of this study are valid only insofar as the protocol was interpreted uniformly by all investigators throughout the study. Meetings of the principal investigators and statisticians were held about every two months and sometimes more often. In addition, telephone conferences were frequently used for consultation about procedure and interpretation. These factors should not be overlooked in the budgetary planning of cooperative clinical trials.[16]

So began the collaboration that would eventually grow into the first cooperative clinical trials group network. Due to his seniority, the Acute Leukemia Group A was established under the Sloan-Kettering physician-researcher, Joseph Burchenal, while Acute Leukemia Group B was the title given to Frei's group (within a few years, Study Group A had turned its attentions to solid tumours in children, and the group was renamed as the Children's Cancer Group).[17]

When Zubrod was promoted to Scientific Director of the NCI in 1961 he created the organization's first 'task force' to, as he put it, 'engineer the cure of acute leukaemia'.[18] The resulting Acute Leukemia Task Force brought together members of the leukaemia cooperative groups A and B as well as a number of other academic and industrial scientists in the US and abroad. The 'task force' concept itself was one that Zubrod had borrowed from industry, specifically the computer giant IBM. For Zubrod the task force was the means thorough which all stakeholders in a project could be encouraged to design and implement common standards and to share ideas freely between different institutional and disciplinary groups. It was a style of research and management that he would extend to other cancers during the rest of his tenure at the NCI.

The defining early success of the Leukemia Task Force, a success that would help secure the reputation of the cooperative clinical trial as a viable and valuable tool, began in 1960 when the pharmaceutical house Eli Lilly made their new anticancer drug Vincristine available to a handful of researchers at the NCI. Emil Frei and his colleagues at the NCI's Clinical Center, Emil Freireich and Myron Karon successfully used it (in high doses) to induce an unprecedented remission in children suffering from acute lymphocytic leukaemia. Following discussion with their task force colleagues, Frei and Freireich designed a trial for Vincristine to be used in conjunction with other chemotherapies—the so-called VAMP regimen—the results of which were impressive.[19] Following the successes of VAMP, Zubrod's recounts that it was a much easier sell to persuade Congress and professionals alike that task forces be formed for virtually every type of cancer.[20]

One key logistical triumph of the cooperative programme was their ability to (at least in part) overcome some of the difficulties in recruiting and retaining patients with relatively rare diseases (like children with leukaemia, for instance). Another feature of the system, though, explains why the technique took hold in the more commonly seen diseases. It was becoming obvious by the early-to-mid 1960s that the striking and clearcut results often seen in the early tuberculosis trials were the exceptions, not the rule. It followed that if many drugs were likely to show relatively marginal activity, the ability to test across large populations became desirable in order to better determine statistical differences. The Director of the NCI during the 1960s, Kenneth Endicott, said of this:

> Specific measures are taken to secure uniformity, such as frequent meetings and inspection of individual laboratories by each member of the group. These restrictions would not be necessary if the treatment produced such dramatic effects as penicillin does in lobar pneumonia. Unfortunately such effects on cancer are the exception rather than the rule.[21]

The solution to these problems was then the so-called 'cooperative (or multi-center) clinical trials' programme that used central organizing, administrative bodies linking multiple clinical locations all of which shared a trial design or 'protocol'. The Southwest Cancer Chemotherapy Study Group (later renamed the Southwest Oncology Group, or SWOG) founded in 1956 and headquartered in Houston, Texas was an early example of such an organization, its members being strongly represented within the Acute Leukemia Task Force and the will to come together to pool expertise and scarce paediatric patients.

The view I have sketched out here of American medical research changing so much so fast is not the whole picture, however. Although the 'establishment' elites of academic medicine were increasing in their influence in the years after WWII, they were not unopposed in their efforts by other academics and clinicians. The Wooldridge Committee appointed by President Lyndon Johnson in 1964, for instance, was quite critical of aspects of the NCI, most particularly the accountability and management of the CCNSC.[22] By the mid-1960s, the NIH was disbursing something of the order of $1 billion a year, making oversight an increasingly pressing issue. Zubrod, in hindsight at least, took the criticism in his stride, dismissing much of it as backward looking: 'most of us [at the NCI] concluded the Wooldridge Report was a symptom of relapse of the chronic conviction of the scientific community that the engineered approach to biomedicine is necessarily bad'.[23] In their account of the NCI's early leukaemia efforts the social scientists Peter Keating and Alberto Cambrosio, seem to share Zubrod's belief that development of cancer chemotherapy trials was somewhat inevitable and when in operation, overwhelmingly convincing. The very nature of the task force, they argue, transformed what was originally designed to be a test of drugs into a fundamental inquiry into disease.[24] The clinic produced ideas for study just as surely as the bench-side did, but under the clinical trial system those ideas had a much better chance of being disseminated beyond the confines of isolated research groups. As I will discuss in Chap. 6, RCTs were and still are somewhat more controversial and messy than this. This quest for uniformity proved to be, in reality, a difficult thing as different groups of researchers 'tweaked' and modified clinical trial protocols within their own institutions. Nevertheless, the *idea* of uniformity, of 'clean' data, helped clinical trials become (and stay) the gold standard of clinical research to the present day in spite of the flaws inherent when careful planning meets individual idiosyncrasy. Clinical trials were crucial in promoting the chemical approach as the cutting edge of cancer care, nudging aside both surgery and radiotherapy in terms of prestige and research dollars. By the time of the Wooldridge Report, the CCNSC was swallowing up some $47.5 of the $58 million dollars a year set aside for collaborative research at the NIH, something that for critics and members of the Committee alike felt to be a case where 'availability of money exceeded the availability of sound ideas', which led to a public bashing of the NCI in general and the CCNSC in particular.[25]

For the first few decades of their existence, then, cooperative clinical trials largely focused on drug treatments to treat diseases. Even in 'multimodal' (that is, using a combination of treatment modalities) and 'multidisciplinary trials', surgeons and radiation oncologists tended (in the US at least) to be pushed to the margins in terms of input into protocol design and management, certainly until the later 1970s. How surgeons and radiotherapists responded to this state of affairs is discussed in Chap. 6, but for the reminder of this chapter I return to the prostate and the status of prostate cancer as an object of the new cooperative clinical trials.

THE VETERANS ADMINISTRATION AND COOPERATIVE CLINICAL TRIALS

In 1924 an act of the Congress made it a right of veterans of any US war to access a hospital bed regardless of how the injury or sickness that put them there occurred. Up to that time veterans had had some access to a patchwork of sickness homes and pensions, but the act of 1924 helped the then Veterans Bureau (later Veterans Administration or VA) create a new system of hospitals and medical services. What the architects of the law could not know was how, over time, the VA hospitals would go on to radically shape the growth of academic medicine in the United States. Real change began to occur in the aftermath of WWII when President Harry Truman signed Public Law 79-293 authorizing the VA to establish a Department of Medicine and Surgery with a broad remit to run education and research programmes. A temporary 'bump' in patient numbers was anticipated by the military as soldiers returned from the war, and the medical schools provided welcome staffing resources to VA wards and outpatient clinics. Similarly, as a result of the GI Bill many veterans were expected to return from the war and use their tuition money to train in the professions, including medicine, so the additional residency capacity created was welcomed by the medical schools also. The postwar surge of patients and medical students was exactly what *did* occur, but in the years that followed the VA hospitals and the institutions of academic medicine became more and more entangled in ways that made a temporary situation transition to a state of permanency.

There are several reasons why this coupling became so hard to undo. The policy of the VA to build its hospitals close to medical schools certainly helped to meld the different institutions, as did the vast new grant monies that the NIH made available to researchers at those schools if they could secure a clinical population on which to conduct investigations. With the rise of private health insurance during the 1950s, the hospitals were increasingly able to admit fee-paying patients over charity cases. The ability of these new 'consumer' patients to push back against intrusive educational and research interactions led to problems for the academic departments, quite used as they were to free access to the bedsides of a largely compliant patient population. Within twenty-five years of the signing of Public Law 79-293 there were over a hundred VA hospitals, and of the one hundred and twenty medical schools then in existence almost ninety-five per cent of them had educational and research connections with those federal hospitals.[26] For its part, the VA received vital clinical support with an estimated seventy-seven per cent of its beds and eighty-one per cent of its patients covered by students, residents, and senior attending staff from medical schools and surgical departments of university hospitals.[27] By the early 1980s, some fifty per cent of all physicians and surgeons practising in the US had received at least some of their education and training within the VA system.[28]

The close relationship between the VA and medical schools also meant that the new trends in academic medicine were bound to percolate into the VA. Indeed, when the RCT first emerged onto the international research scene in the 1940s the VA was an early adopter of the model. In one important way it was obvious that the VA would look at the streptomycin study in the UK with considerable interest—tuberculosis had, after all, long been the scourge of the military, and the hospitals of the VA were set to accept returning veterans, a portion of whom would likely be infected. They soon conducted their own extensive studies into the new chemotherapy agents.[29] That said, the origins of the VA's first RCTs for tuberculosis also lie in the rationalization of the United States Public Health Service (USPHS) during and after WWII. Before the war the USPHS had a confusing array of research and social service duties that, as the great chronicler of American medicine Paul Starr put it, 'reflected the diverse repertoire of bit parts the federal government was called upon to play in medicine'.[30] Just when the federal government was about to take a much larger role in health research through the expansion of the NIH, a 1944 act of Congress harmonized the functions of USPHS with those of the NIH. As part of this reorganization, the Surgeon General convened the three national advisory councils—the National Advisory Health Council, the National Advisory Cancer Council, and the National Advisory Mental Health Council—and mandated them to organize a variety of 'study sections'. The resultant sections were mostly organized into specialty areas, like pathology or gerontology, but some were also focused on particular diseases such as malaria or tuberculosis.[31] As the then Chief of the USPHS Research Projects Gordon Seger said of this in a 1947 meeting:

> With the aid of the members of the Study Sections and the three advisory councils, comprising some 250 of the leading scientists in the nation, a large-scale, nationwide, peacetime program of support for scientific research in medical and related fields has been in operation for almost two years. The underlying philosophy of this program is predicated on the complete autonomy of participating investigators. However, in certain cooperative projects, wherein a number of investigators are engaged in solving a specific problem common to all, such as in the Syphilis and Tuberculosis fields, complete autonomy for each investigator is not always practical. Nevertheless, I should like to emphasize that, in so far as the Federal Government is concerned, this program, based on research grants financed by public funds for the support of research by independent scientists, is devoid of governmental control.[32]

Although expressed in rather less bombastic words than those chosen by Zubrod, here is another recognition that the 'engineered' approach to medical research was far from universally accepted. While many academic researchers welcomed more federal dollars for research, as the historian of bioethics Laura Starks points out, they were much less happy about the need to 'hitch themselves to a specific research mission'.[33] Federal administrators would face another, related, challenge after WWII when the American Medical Association (AMA) looked askance at the nationalization of the health services then under way in Great Britain and began to campaign quite vigorously against what they deemed to be a form of 'socialized medicine'. As Stark and others have argued, with a war against totalitarianism just fought and won, the AMA sought to bolster its opposition to health reform by linking government involvement in healthcare to the atrocities of Nazi Germany; so it is understandable that Seger should have sounded so cautious in this climate of paranoia.

In spite of these fears of government intrusion into science and medicine, in 1946 the Tuberculosis Study Section was founded by a one million dollar appropriation from Congress to begin trials of streptomycin. The RCTs were, though, intended to be a part of a larger, comprehensive programme designed to, 'not only the testing of a single therapeutic agent, but [to] provide plans for a more comprehensive search for the most effective therapy, including the development, the laboratory and animal testing, and the clinical trials of any antibiotic or chemical compound which might offer promise'.[34] For the trials part, the studies were to be organized across institutions and centrally planned as to ensure 'the collection of uniform observations that may be combined or pooled to furnish statistically significant evidence as to the value of streptomycin in the treatment of certain well-defined types of pulmonary tuberculosis'.[35] The trials were deemed to be a great success and the model was soon adopted for other, non-communicable diseases including cancer. By the late-1950s the CCNSC, had organized around ten cooperative groups working in roughly one hundred hospitals, many of which were operated by the VA.[36] Direct NCI support for this research was quite short-lived. Whereas other groups supported by the NCI focused on childhood cancers and adult leukaemia for the insight into basic biological function that they brought,[37] from the outset the VA programme was different. In focusing on much more demographically prevalent diseases—lung, stomach, colon, and prostate, and the like—the VA research service tackled the pressing problems of their patient population but did so with less hope for the dramatic 'chemical cures' seemingly so tantalizingly in reach of the other early NCI cooperative groups.[38]

One of the hospitals to engage in this new research culture was the VA medical center in Minneapolis, Minnesota. It was here one day in late 1950 that a recently discharged captain from the US Army Medical Corps Reserve—a man whose name would soon become part of the lexicon of prostate cancer and remain so to the present day—would report for duty to take up the post of Chief of Pathology: Donald Gleason. Before arriving in the Midwest, Gleason and his wife had lived in Paris, drawn there by the art and culture of the city, and while in France Gleason indulged in his passion for sketching and drawing so developing a skill that would become of crucial importance to his career as a pathologist within the VA system.[39] Another and perhaps more predictable element in Gleason's rise to prominence was the vision and collegiality of his boss at the Minneapolis VA, George Mellinger, Chief of Urology. Mellinger was a great student and enthusiast of the work of Charles Huggins on the hormonal control of prostate cancer (see Chap. 4), and the urologist saw the RCT as the ideal means through which to test the efficacy of anti-androgen treatments.[40]

An immediate problem Mellinger faced in trying to organize such trials, though, was exactly the lack of objective measures that had so bedeviled Huggins in his research. While phosphatase levels provided some information, they could not hope to stand in for the kinds of sophisticated tools used by the Tuberculosis Study Section to track pulmonary tuberculosis—x-ray images of the lungs, bacteriological blood analysis, and the like. The historian of biomedicine, Ilana Löwy, points to what would become a critical turning point in cancer clinical trials in general, and the story of Mellinger and Gleason in particular—the development of 'cancer staging', a system of defining the severity of malignant disease on the basis of clinical and pathological findings.[41]

THE EMERGENCE OF THE 'GLEASON SCORE' AS A PROGNOSTIC INDICATOR

In 1954 Mellinger began to organize a cooperative group to study carcinoma of the prostate in (mostly) veterans of WWI who were by then high users of the wards and outpatient clinics of the VA system. The group he created, the Veterans Administration Cooperative Urological Research Group (VACURG), included only those VA medical centers with a full-time urologist on staff, which at that time represented some fourteen hospitals across thirteen states. Closer to home, Mellinger turned to his colleague Gleason to create some clear pathological indicators to support VACURG in their work. As Gleason said of this in his seminal 1966 paper (quoted in the epigram to this chapter):

> Carcinoma of the prostate shows great variation in its natural history and also has a wide range of histological characteristics. Various systems of histologic grading of carcinoma of the prostate have shown varying degrees of correlation with the apparent degree of malignancy of the cancer, as measured by such criteria as survival time or presence of metastases. All of these

classifications are necessarily subjective and it is difficult to know if one is following another author's classification accurately.[42]

In their biographical study of Gleason, his former colleagues, John Phillips and Akhouri Sinha, argue that such a quest for uniformity was enormously appealing to the pathologist given his earlier work as a student researcher on the Minnesota Multiphasic Personality Index (MMPI) in the 1940s. The MMPI used logistic regression—a type of mathematical modeling—to develop a standardized system of psychiatric testing and diagnosis. It was work that convinced Gleason that the practice of medicine was ripe for the kinds of optimization that statistical analysis could supply. Phillips and Sinha quote Gleason as saying something of this in a later reflection:

> It was obvious that pathologists had difficulty adopting (different) classifications. A few photomicrographs and thousands of words did not serve to convey them to others ...

> Thanks to an experience while a medical student ... watching the development of the MMPI, I felt that I could do a better job than the prevailing state of affairs. I was asked by Dr. George Mellinger to devise a grading system ... and I agreed to try.[43]

As was the case for researchers putting together the MMPI, Gleason decided that in order to devise a 'reference standard' he would need to put aside conventional wisdom, in his case on the subject of prostate cancer, and instead look with fresh eyes for novel patterns:

> I felt that the way to develop a histologic classification was to forget anything I thought I knew about the behavior of prostate cancer and simply look for different histological pictures. ... Then, the pictures would be handed to statisticians and compared with a 'gold standard' of clinical tumor behavior (i.e., patient survival).[44]

The 'pictures' Gleason talked about were actually the two hundred and seventy histological specimens he reviewed for Mellinger, all of which were taken from prostate cancer patients at the Minnesota VA hospital. These were patients that Mellinger had first selected and then randomized into the VACURG's first study, a prospective trial comparing different treatments including patients 'chemically castrated' via Huggins' hormone method (see Chap. 4).

Before being admitted to the first trials of the VACURG, first hundreds and then thousands of patients (eventually over five thousand in all[45]) were clinically 'staged'. The criteria for prostate cancer stages I–IV that the VACURG used was first laid out by Willet Whitmore in the Memorial Center for Cancer and Allied Diseases in the 1950s (the numerals were later replaced with the letters A–D to improve clarity when clinical staging was combined with numerical grading information).[46] Under Whitmore's system, 'stage I' represented an 'incidentally' found microscopic cancer (due to removal of prostate tissue for 'nonmalignant' disease of the prostate); 'stage II', palpable nodule; 'stage III', local spread (determined via palpation of contiguous tissue and lymph nodes); and 'stage IV', distant metastases with or without elevated phosphatase levels (determined using x-rays for bony metastases or by biopsy of distant lymph nodes).[47] This clinical staging was somewhat imprecise but it did allow VACURG to do the work of dispersing patients across different arms of the trial and then also gave Gleason an index with which to compare his new method of pathological grading.

In beginning work on this new, as yet unspecified, method of grading, none of the clinical staging information was made available to Gleason— he initially received the specimens without any associated clinical data or identifier apart from a simple slide number. To add to the mystery, Gleason did not even know if each of the slides represented a unique patient or if some were multiple samples from a tumour in any one patient.[48] In Gleason's mind, nine distinct 'patterns' emerged ranging all the way from well differentiated, organized, and uniform glands to a state where glands showed very poor differentiation and had extensive infiltration of the stroma (connective tissue) of the prostate. Gleason sent off his findings to a team of statisticians at the NCI headed by John Bailar. Mellinger, meanwhile, had 'deanonymized' the samples, supplying other data for Bailar including information about the subsequent clinical course of patients represented in the samples, as well as specific information on the kinds of specimens represented on each slide (from instance, whether they were taken from biopsies or from open prostatectomies).[49] After a preliminary work through of the trial materials, Bailar's team decided to focus on two main data points: the combined score of the two patterns predominating in any one sample, and the clinical stage I–IV. Bailar then compared those data to the survival outcome of individual patients. As Gleason would later recall about the work of the statisticians: '[They] found surprisingly strong correlations between my histologic pictures and the patient death rates and cancer-specific death rates. They recombined my nine pictures into five "patterns" because those combined patters had very similar mortality rates.'[50]

Gleason and his collaborators pressed ahead with devising a simple numerical method of making a prognostic 'score' based on these five patterns. As Gleason had determined that it was usually the case that more than one histological pattern appeared in any one specimen, the researchers took both the predominant and the second most common pattern to calculate the first part of the score. The second part of the score was the numerical value of the clinical stage of the cancer. The lowest possible theoretical score was a 3, meaning that the two most prevalent histological patterns were each graded at 1, and the patient presented at clinical stage I (so 1 + 1 + 1). Similarly, the theoretically worst possible prognostic score was a 14, for a patient demonstrating a clinical stage IV cancer and 5 + 5 grading on histology. After being rejected by a couple of urological journals,[51] Gleason finally published his new scoring system in his 1966 paper in the NCI's *Cancer Chemotherapy Reports* as sole author. As for the collaborators, Bailar reported as first author on his statistical techniques in the same issue of the journal, and a year later Mellinger published as first author in the *Journal of Urology* discussing the predictive value of the technique. The results thus gained good coverage in both 'old' style surgical communications and the new style of cancer research.

This is not to say that the Gleason method was the only system of prostate cancer grading circulating in the academic community at this time. Such was the need for clarity on the relative benefits of the competing methods that in 1979 the American Cancer Society sponsored a series of 'consensus workshops' bringing together clinicians, pathologists, and statisticians to consider and hopefully determine the relative merits of the different techniques. They considered the Gleason, Mostofi (developed at the Armed Forces Institute of Pathology), Gaeta (developed at Roswell Park Cancer Institute), and the Mayo Clinic systems side-by-side. The final report commended the Gleason system above others, but not strongly so:

> Experience within and outside of the workshops has demonstrated that the Gleason system is quite readily learned and reasonably reproducible. Although Dr. Gleason estimates his own reproducibility rate to be 80 %, in studies by others reproducibility was approximately 70 %. The margin of

error of reproducibility from one institution to another could be as much as 50 % and probably reflects the degree of experience and instruction given to the particular observer.

Although clinical staging information has been utilized in conjunction with the Gleason system in defining patient groups which correlate remarkably well with patient survival, it was generally agreed that such combinations of histologic classification and clinical staging should not generally be adopted in evaluation of clinical grading systems since criteria of clinical staging vary in different places and at different times and since most reported systems of histologic grading have not incorporated such analyses.[52]

Whitmore himself had recognized in his original 1956 paper on the clinical staging of prostate cancer that his attempts to impose order to better research a baffling disease were beset by problems:

> With regard to this method of classification into stages it is important at the outset to point out the following facts: 1. The classification is arbitrary since the cancer process is a dynamic one whereas the stages, as defined, are static. 2. The classification is clinical rather than pathologic. The significance of this limitation is due to the fact that the pathologic stage of the disease may be different from the clinical stage. Stage I cancers are not always small. Stage II cancers are not always pathologically confined within the prostatic capsule. This is abundantly clear from the therapeutic failures following attempts at cure of stage II lesions with total excision by means of radical prostatectomy.[53]

Whitmore saw the lack of understanding of tumour progression as exacerbated by the wide variation in how tumours behaved in different patients:

> Doing what one believes to be right provides the moral justification for therapeutic excursions into the valley between these two mountains of ignorance, and what one believes right may, depending upon the circumstances, vary from virtual therapeutic nihilism at one extreme to the most radical therapeutic efforts at the other.[54]

As limited as Whitmore himself saw his system, he nonetheless saw it as crucial for mounting the kinds of large cooperative studies advocated first by Nesbit and Baum in 1950 and actually carried out by VACURG in the 1960s and 1970s (as well as trials conducted in the 1970s and 1980s by the National Prostatic Cancer Project described below). The Gleason and

Whitmore classification schemes both underwent significant revisions in 1974[55] and 1975[56] respectively, but the shift to Gleason as the dominant system was, as evidenced by the 1979 conference, a slow one, occurring over a decade or more.

The 1979 consensus conference did not dismiss the other systems out of hand, instead the suggestion was that these tools might somehow work 'in conjunction' with the Gleason score. This was possibly a mollifying kind of assertion, one that was perhaps mindful of the prestige of the academics and institutions involved in designing the 'lesser' classifications (the author of the published report, Gerald Murphy of Roswell Park, was himself unlikely to be a disinterested voice in this regard), but it was not a position that provided particularly clear advice for ordinary clinicians.[57] The conference did have other reasons to be conservative in its recommendations in addition to avoiding ruffled feathers and the stated concerns over Whitmore's clinical staging criteria. The conclusion of the report touched again on broader issues of the reproducibility (and so reliability) of the methods of tumour grading:

> All classification systems have to deal with a number of basic considerations: 1. the tissue available for sampling, 2. the objective definition of grading criteria, 3. the degree of reproducibility of interpretation, 4. simplicity, and 5. the predictive value of the system relative to the biologic potential of the tumor. All of the systems discussed met these requirements to a relatively similar degree. With reference to the biologic potential of tumors, it was pointed out that ... at the present level of knowledge grading systems do not reliably predict the lethal potential of a tumor in an individual patient nor the responsiveness of an individual tumor to various forms of therapy. Caution was repeatedly expressed regarding the use of tumor grade in the individual patient as a basis for treatment decision.[58]

Given this uncertainty then, it might seem more appropriate to ask why the Gleason system *did* (rather than did not) emerge as the consensus choice, something that the published report is remarkably silent on.

Phillips and Sinha make a strong case for the success of Gleason grading as being related to Gleason's days in France where he developed is eye for artistic representation. While the other systems rendered their different categories using charts and tables, Gleason created something visually arresting and compelling in its accessibility and elegance—a graduated pictorial illustration of tumour grading. Of his original hand drawing (which he had created in preference to showing a series of photographic images of slides), Gleason said: '[it was] probably my most valuable contribution to the grading system. Pathologists, who tend to think in pictures, quickly grasped this graphic representation'.[59] Gleason's modesty aside, and as visually arresting as the drawing was and iconic as it quickly became, this explanation is probably not enough to explain the persuasiveness of his system. I think the other part of the explanation probably lies in the fact that it was not just pathologists to whom the system had to 'speak'. The consensus conference was, after all, designed to be multidisciplinary and clinicians, specifically urologists and radiologists (whose role is discussed in Chap. 7), biostatisticians and other kinds of (non-pathologist) scientists were also strongly represented.[60] Unlike the other tumour grading systems, which depended on detailed cytological analysis (that is, they considered both glands *and* cells), the Gleason score was a matter of defining the state of glandular 'architecture'. This disordered architecture could be demonstrated relatively straightforwardly in a prostate sample using a low-magnification microscope (even if those same skilled pathologists might ultimately disagree with each other on what the final 'score' ought to be). As such, this ease of demonstration created a shared 'language' with which pathologists could talk to clinicians and through which both could talk to biostatisticians. Without the additional intricate cytological descriptions of the cells themselves—how they were dividing and aging, for instance—Gleason-graded tumour data was (following instruction by Mellinger on the issue) by design user-friendly in mass cooperative clinical trial scenarios. As a later review of Gleason's system would point out: 'It was recognized early in the VACURG study that tumour grading would be a large undertaking, and as a matter of expediency it was agreed that any proposed grading system would be based upon the general histological appearance of the tumour, rather than relying on specific counts of mitoses or cell types.'[61] The success of the Gleason score shows us that organizational demands of large-scale clinical research not only shaped how institutions and practitioner networks formed and developed, but also such 'macro' concerns could ultimately shape how researchers thought about the world all the way down to the cellular level. The Gleason score was an expedient tool, but one that itself created new questions about clinical and basic research, and so, in turn, affected the macro-level concerns that had first helped to create it.

An example of this came in the 1980s when prostate cancer was incorporated into the 'TNM' (Tumour-Node-Metastases) system developed by the American Joint Committee on Cancer (AJCC) and the Union Internationale Contre le Cancer (Union for International Cancer Control, UICC). TNM classification incorporated the extent of primary tumour (T-category), regional lymph node involvement (N-category), and the presence or absence of distant metastases (M-category) and so provided further precision and reliability in the support of treatment decisions. The TNM came to increasingly replace the Whitmore system as a prognostic indicator, with the revised Gleason score persisting alongside of it. By the turn of the twentieth century the Gleason score was recognized as a category one prognostic parameter by the College of American Pathologists, and was further endorsed by the World Health Organization (WHO) and the AJCC-UICC.[62] The realization of the Gleason score as part of routine practice was therefore not the result of some overnight sensation but was instead the result of a decades long journey through the new institutions and cultures of international biomedicine.

THE 1971 NATIONAL CANCER ACT AND PROSTATE CANCER

The NCI established a Breast Cancer Task Force in 1967 but it took the passage of the National Cancer Act in 1971 for similar groups to be organized around other common solid tumours (see Chap. 6 for a discussion of the politics of breast cancer at this time). The National Organ Site Programs Branch of the NCI oversaw the creation of task forces in bladder, large bowel, and prostate cancer (with lung cancer being split between the Tobacco Working Group and the Division of Cancer Cause and Prevention). While the task force for breast cancer stayed headquartered at the NCI in Bethesda, the other programmes moved out to some of the country's most prominent cancer research hospitals: bladder to Gilbert Friedell at St. Vincent's Hospital, Massachusetts; large bowel to Murray Copeland's team at the M.D. Anderson Hospital and Tumor Institute in Houston, Texas; and prostate to Gerald Murphy's department of urology at the Roswell Park Memorial Institute, New York. This decision to locate the task forces outside of the Bethesda seems to have been inspired in part by criticisms that other NCI targeted research programmes were overly rigid and inflexible, unable to nimbly respond to unexpected findings.[63]

The organ task forces shared some similarities and differences with earlier federal disease-focused programmes of the immediate post-WWII era. Whereas grants submitted to the NCI contracts programme were subject to robust oversight from salaried government scientists, grants awarded to the prostate, bladder, and bowel task forces relied on more traditional sources of academic scrutiny, especially peer-review of grants. While the requirements to follow preplanned programmes of research may have loosened somewhat because of this, some similarities did persist between the 1970s organ group and the 1940s-era task forces in that both kinds of programme were charged with an ambitiously wide remit. In the case of organ groups, this remit stretched from inquiries into epidemiological assessment of the incidence and prevalence of specific cancers, through to screening and treatment studies. For Murphy's team in New York, fulfilling this mandate involved working with, and trying to supplement, the efforts of the VA. Given the overall chemical bent of NCI programmes it made sense that the new prostate task force would—unlike the VA—at first focus on chemotherapeutic approaches to therapy.

In the decade and a half or so of its existence, Murphy's self-styled National Prostatic Cancer Institute and its associated National Prostatic Cancer Project (NPCP) headquartered at the Roswell Park Institute, pursued a largely chemical (and later radiological) agenda. From 1972 to 1985 the NPCP conducted twenty-four trials assessing the use of chemotherapies, and later radiotherapies, as secondary or 'adjuvant' interventions intended to supplement primary treatment with surgery (and) or hormone interventions.[64] That the NPCP was such a short-lived enterprise owes much to the massive reconstruction NCI programmes during the early-to-mid 1980s. Criticism of the management of the NCI, and indeed ongoing criticism of the very policy of planned research (however defined) remained fierce throughout the 1970s. In fact this animosity only took an upturn after the NCI was awarded more resources as a result of the 1971 Cancer Act, when some scientists were very vocal in publically denouncing the division of resources which they felt unfairly deprived other institutes of the NIH of their fair share.[65] Tensions would not improve when budgetary shortfalls across the NIH in the early 1980s affected all programmes, including those of the NCI. Following from recommendations issued by the National Cancer Advisory Board, the NCI reorganized its Organ Site Program into a new format retitled as the Organ Systems Program. The creation of a single headquarters (the Organ Systems Program Coordinating

Center housed at Roswell Park) meant that the NCI could economize on coordination and administration, but once again the move generated considerable controversy for the beleaguered leaders of the institute, including its then director Vincent DeVita (quoted in the opening of this chapter).[66] The restructuring also saw the dissolution of the individual programmes, like NCPC, and their integration into one of the two new branches—gastrointestinal and genitourinary—of the Organ Systems Program.

The clinical trials carried out by the NCPC had not done a great deal to advance the treatment of prostate cancer. While chemotherapy could help some patients whose tumours were resistant to hormone therapy, the fraction of patients helped in this way was comparatively small, and the relief they experienced was temporary.[67] These problems aside, along with the VA trials, the work of the NCPC did succeed in creating a network of knowledge and expertise in cooperative clinical research that would influence a generation of urologists. In time, as I will discuss in Chap. 6, the whole issue of clinical trials would start to become very vexed indeed and the study of prostate cancer became deeply embroiled in these controversies.

THE SIGNIFICANCE AND AFTERMATH OF THE VACURG TRIALS: BEYOND THE GLEASON SCORE

The results of the first VACURG clinical trial (along with two follow-up ones) showed that despite some serious toxic effects hormone treatment for prostate cancer could indeed act as Huggins had promised—relieving symptoms and prolonging lives.[68] While the results may have disappointed from the perspective of advancing a hormonal 'cure' for cancer, they were hugely influential in showing the palliative value of endocrine treatments in patients with advanced disease. Moreover, by the time these trials were under way fresh work was coming out of the laboratory that would drastically improve outcomes in the clinic. Adding to an already remarkable fifty-year chronology for the 'organotherapy' story (described in Chap. 5), work by the endocrinologist Andrew Schally in the 1960s and 1970s (aimed at developing drugs for both infertility treatments and for contraceptive purposes) had lead him to the isolation and characterization of gonadotropin-releasing hormone (GnRH). It was an accomplishment that would bring—in 1977—yet another Nobel Prize in

Medicine or Physiology to the field of endocrinology, but it was also a finding that helped create a whole new class of drugs able to reduce the tumour-promoting effects of testosterone in the body of the prostate cancer patient, offering more palatable alternatives to existing hormonal and surgical castration treatments.

The two-way flow of ideas and data between the bench-side and the bedside in prostate cancer is certainly as interesting as it is evident. The data gained from mass trials made it possible for clinical knowledge to meaningfully inform research in the laboratory and the clinic. This concept of reciprocity was of critical importance to the success and growth of clinical trials in the latter twentieth century more generally. The federal Office of Technology Assessment pithily summarized the significance of this principle in its 1983 report, *The Impact of Randomized Clinical Trials on Health Policy and Medical Practice*:

> In a broader sense, RCTs can be used to answer questions susceptible to the scientific method about interventions involving human beings. Well-designed and executed RCTs are not merely product testing, but should answer questions about important hypotheses. They should, therefore, generate biologically and medically important information.[69]

One obvious example of how such reciprocity worked in the case of prostate cancer was the clinical, pathological, and statistical creation of the standardization of tumour grading that became known as the Gleason score. While the standardization allowed trial data to be more rigorously analysed, the trials, in turn, provided an abundance of data on the shortcoming of, and suggested improvements to, the scoring system. As in most large clinical trials, then as now, answers were not clear cut but they did provide data for experts to gather in consensus conferences like the ones sponsored by the American Cancer Society in 1979. This kind of work was probably trials at their most potent, but as we will see in the following chapters, the status of RCTs as the gold standard of clinical research, while never free from controversy, became increasingly contested by the close of the twentieth century. When a third prognostic indicator for prostate cancer—the prostate specific antigen, or PSA test—popped up in the late 1980s, a new kind of problem emerged as the number of men diagnosed with prostate cancer began to soar; a kind of problem that would embroil old arguments about clinical trials and strike at the very notion of what 'evidence' meant in medicine.

NOTES

1. Gleason, Classification of Prostatic Carcinomas, 125.
2. DeVita, The Governance of Science at the National Cancer Institute, 3973.
3. Medical Research Council, Streptomycin Treatment of Pulmonary Tuberculosis.
4. Yoshioka, Use of Randomisation in the Medical Research Council's Clinical Trial of Streptomycin in Pulmonary Tuberculosis in the 1940s.
5. Chalmers and Clarke, Commentary.
6. Marks, *The Progress of Experiment*.
7. Kutcher, *Contested Medicine Cancer Research and the Military*, 26.
8. Gilman, Therapeutic Applications of Chemical Warfare Agents.
9. Bud, Strategy in American Cancer Research after World War II, 433.
10. Johnstone and Baines, *The Changing Faces of Childhood Cancer*, 48.
11. Farber et al., Temporary Remissions in Acute Leukemia in Children.
12. The term 'chemotherapy' was first coined by the German physician and scientist Paul Ehrlich after his laboratory went in search of an arsenical compound that might be biologicaly active against syphilis, and found one in 'Number 606' of his series. Arsenic was at that time routinely administered to patients suffering from syphilis, but the production of Number 606, or Salvarsan as it came to be called, allowed for higher doses of the active arsenical agent to be administered without the associated higher toxic effects (an approach Ehrlich termed the 'magic bullet'). See Parascandola, The Theoretical Basis of Paul Ehrlich's Chemotherapy.
13. Zubrod, Origins and Development of Chemotherapy Research at the National Cancer Institute, 12.
14. Zubrod, Origins and Development of Chemotherapy Research at the National Cancer Institute.
15. Frei et al., A Comparative Study of Two Regimens of Combination Chemotherapy in Acute Leukemia.
16. Ibid., 1142.
17. Zubrod, Origins and Development of Chemotherapy Research at the National Cancer Institute, 11.
18. Ibid., 14.
19. Ibid., 13–15.
20. Ibid., 16.
21. Endicott, The Chemotherapy Program, 285.
22. Cooper, Onward the Management of Science.
23. Zubrod, Origins and Development of Chemotherapy Research, 17.
24. Keating and Cambrosio, From Screening to Clinical Research, 305.
25. Cooper, Onward the Management of Science, 1436.
26. Gurll et al., The Veterans Administration and Academic Surgery, 2.

27. Ibid., 4.
28. Gronvall, The VA's Affiliation with Academic Medicine, 63.
29. *Transactions of the Streptomycin Conference.*
30. Starr, *The Social Transformation of American Medicine*, 342.
31. Seger, A Cooperative Study of Streptomycin in Tuberculosis, 686–7.
32. Ibid., 687–8.
33. Stark, *Behind Closed Doors*, 100–101.
34. Seger, A Cooperative Study of Streptomycin in Tuberculosis, 689.
35. Ibid., 690.
36. Löwy, *Between Bench and Bedside*, 55.
37. Johnstone and Baines, *The Changing Faces of Childhood Cancer*, 52.
38. Hrushesky, The Department of Veterans Affairs' Unique Clinical Cancer Research Effort, 2703.
39. Phillips and Sinha, Patterns, Art, and Context, 497.
40. Ibid., 499.
41. Löwy, *Between Bench and Bedside*, 55–6.
42. Gleason, Classification of Prostatic Carcinomas, 125.
43. Phillips and Sinha, Patterns, Art, and Context, 499.
44. Ibid., 500.
45. Gleason, Histologic Grading of Prostate Cancer, 273.
46. Whitmore, Hormone Therapy in Prostatic Cancer.
47. Mellinger, Gleason, and Bailar, The Histology and Prognosis of Prostatic Cancer, 332–3.
48. Gleason, Classification of Prostatic Carcinomas, 125.
49. Mellinger, Gleason, and Bailar, The Histology and Prognosis of Prostatic Cancer, 331.
50. Phillips and Sinha, Patterns, Art, and Context, 501.
51. Ibid.
52. Murphy and Whitmore, A Report of the Workshops on the Current Status of the Histologic Grading of Prostate Cancer, 1490–1.
53. Whitmore, Hormone Therapy in Prostatic Cancer, 698.
54. Ibid., 703.
55. Gleason and Mellinger, Prediction of Prognosis for Prostatic Adenocarcinoma by Combined Histological Grading and Clinical Staging.
56. Jewett, The Present Status of Radical Prostatectomy for Stages A and B Prostatic Cance.
57. Murphy and Whitmore, A Report of the Workshops on the Current Status of the Histologic Grading of Prostate Cancer, 1493.
58. Ibid., 1492–3.
59. Phillips and Sinha, Patterns, Art, and Context, 501.
60. Murphy and Whitmore, A Report of the Workshops on the Current Status of the Histologic Grading of Prostate Cancer, 1491.

61. Delahunt et al., Gleason Grading, 76.
62. Srigley et al., Prognostic and Predictive Factors in Prostate Cancer, 10.
63. Maugh, *Seeds of Destruction*.
64. Murphy, Review of Phase II Hormone Refractory Prostate Cancer Trials, 19.
65. DeVita, The Governance of Science at the National Cancer Institute, 3969.
66. Ibid., 3972.
67. Huben and Murphy, Prostate Cancer, 287.
68. Gleason, Histologic Grading of Prostate Cancer, 501.
69. Gelband, *The Impact of Randomized Clinical Trials on Health Policy and Medical Practice*, 10.

CHAPTER 6

Screening, Patients, and the Politics of Prevention

The benefit of screening for prostate cancer with serum prostate-specific antigen (PSA) testing, digital rectal examination, or any other screening test is unknown. There has been no comprehensive assessment of the trade-offs between benefits and risks. Despite these uncertainties, PSA screening has been adopted by many patients and physicians in the United States and other countries. The use of PSA testing as a screening tool has increased dramatically since 1988.
    Prostate, Lung, Colorectal, and Ovarian Cancer Screening Trial project (2009)[1]
    I never dreamed that my discovery four decades ago would lead to such a profit-driven public health disaster. The medical community must confront reality and stop the inappropriate use of P.S.A. screening. Doing so would save billions of dollars and rescue millions of men from unnecessary, debilitating treatments.
    Richard Ablin, The Great Prostate Mistake (2010)[2]

The year 1979 was a good one for Gerald Murphy's National Prostatic Cancer Project (NPCP) at Roswell Park. In an article titled Purification of a Human Prostate Specific Antigen, Murphy's team claimed to have purified and identified a new immunological marker linked to prostate cancer,[3] but it was a discovery that a pathologist from the University of Buffalo school of medicine, Richard Ablin, also claimed. Arguments over recognition are, and have long been, a part of doing science. Many researchers in many laboratories do, after all, work on very similar issues so simultaneous discoveries are not uncommon. What made the argu-

© The Editor(s) (if applicable) and The Author(s) 2016   123
H. Valier, *A History of Prostate Cancer*,
DOI 10.1057/978-1-137-56595-2_6

ment regarding the discovery of prostate specific antigen (or PSA) of particular interest though, is that Ablin's results were *not* simultaneous, he had published his data years before and the NPCP team had duly cited it. What's more, in his book, *The Great Prostate Hoax: How Big Medicine Hijacked the PSA Test and Caused a Public Health Disaster* (2014), Ablin reports that he had also applied for a grant (which was rejected) from the NPCP during the late 1970s detailing how he would work on extraction and purification procedures linked to his earlier findings, or, in other words, the work that the Roswell group would go on to publish to considerable fanfare.[4] As PSA-related patents had been filed by the NCPC, lawyers soon became involved, and the dispute rumbled on for years. Again, a distressing set of circumstances for those involved but not unusual within scientific controversies. What really sets this dispute apart, however, is the extent to which it was *retrospectively* ignited when one party—Ablin—subsequently became a highly vocal critic of the use (or as he saw it, abuse) of the very substance he had discovered.

Ablin's researches on the prostate began when he joined the Millard Fillmore Hospital Research Institute (an affiliate institution of the University of Buffalo school of medicine) in 1968 to work with the Institute's two urologists, Ward Soanes and Maurice Gonder. Soanes and Gonder had a major grant to work on the normal and abnormal prostate and Ablin, who had just completed a postdoctoral fellowship in the department of bacteriology and immunology at the school of medicine, assisted in their research of the increasingly popular technique of 'cryosurgery' (the use of intense cold to destroy tissue). Moving between monkey and rabbit models in the laboratory and patients on the wards, the researchers began to observe apparently promising remissions in cases of metastatic prostatic cancer when they used cryosurgery on the prostate.[5] While the original observations of remission were frustratingly elusive in follow up studies, the work did set Ablin on the pathway to discovery. Reasoning that the use of freezing agents had likely induced an immune response leading to the diminution of tumours, Ablin went looking for likely antigen whose release from the prostate might be causing these effects. He looked at the prostatic tissue and secretions from normal, benignly hypertrophic, and malignant human prostates and what he found was that there was indeed a substance (PSA) released from the prostate in various states. He noted that while the presence of PSA in the bloodstream was normal, there appeared to be a marked elevation of PSA in prostatic disease. The basis of Ablin's polemic in the *Great Prostate Hoax* is precisely this: that these observations did not show a *cancer*-specific antigen but rather a *tissue*-specific antigen, there has been, he claims, a woeful and wilful elision between the two ideas in part driven by greed and desire for fame. Ablin has certainly written much more about the discovery of PSA (and his role in that discovery) than any of the other early researchers, and that fact alone makes untangling the claims and counterclaims of PSA difficult. In a 2013 book, *The Prostate Monologues*, the American sports-writer and prostate cancer patient Jack McCallum noted this and described his attempts to reach out to the relevant actors. When he spoke to one of the still living authors (Murphy died in 2000) of the 1979 NPCP study, T. Ming Chu, he found a voice of strong dissent against Ablin's accounts of the dispute.[6]

What even Ablin concedes to be the great achievement of the NCPC team was its transformation of an isolated molecule into a biological marker. That watching this marker could provide a physician with information about how the abnormal prostate was responding to treatment was not and is still not disputed. The idea that grew out of the work at Roswell, though, that such a marker could become a 'test' designed as a *screening* tool applied as part of a routine physical examination of healthy men would go on to create a firestorm of controversy. Ablin directs much ire at the Roswell group for, as he sees it, cynically engineering this step from prognostic to screening tool. (Ablin accuses Murphy of purposefully creating 'a powerful bully pulpit' thanks to political lobbying to ensure that the designation of 'Comprehensive Cancer Cancer' as included in the 1971 National Cancer Act[7] —whatever the veracity of this claim Roswell *was* first centre so designated and we might expect therefore that recommendations on cancer research and screening policy emanating from there carried some considerable weight.) The actions of Murphy, in conjunction with the already inflamed militaristic rhetoric of Nixon's 'war on cancer', Ablin argues, created a gold-rush mentality, and a tendency to seize upon the next 'greatest thing' as a technological arms race in the battle against disease.[8] Add to this already potent mix the involvement of the then nascent biotechnology industry, and we have chain of events that Ablin believed created the 'public health disaster' that he wrote about so vividly and controversially in the 2010 *New York Times*, op-ed.

ROSWELL, HYBRIDTECH, STANFORD, AND THE FDA: BRINGING THE PSA TEST TO MARKET

The science of the PSA test is based upon some remarkable breakthroughs in immunology in the 1970s in the field of monoclonal antibody technology. The detection of an antigen in any biological sample is dependent upon having large amounts of antibody that would specifically bind to it, which in turn rests on an ability to isolate and mass-produce the antibody.[9] By the 1980s one particular biotechnology company based in San Diego, California, called Hybritech was emerging as a leader in the field of monoclonal production and antibody-based diagnostic testing.[10] At a time when new startups in biotechnology were beginning to successfully lure scientists out of the confines of academia with promises of better laboratories and more freedoms, the company, cofounded in 1978 by a Howard Bindorf, a biochemist, and Ivor Royston, a Johns Hopkins-trained physician, was successful in attracting not only talent but also money. Following Roswell's 1979 publication of their work on PSA, Bindorf and Royston reached out to Murphy and his team and began to conduct research and development work on PSA under licence. Recognizing the economic potential of monoclonal antibodies in cancer research, Hybritech drew in first venture capitalists but also the attentions of established pharmaceutical houses.[11] Late in 1985 Eli Lilly bought Hybritech and did so at an extremely auspicious time; the biotechnology company had just developed the first PSA diagnostic test, and, very soon after the merger took place the Food and Drug Administration (FDA) approved its transfer to market as the Hybritech Tandem-R PSA test.[12] This, in Ablin's account, is where things went awry: the FDA had approved the test for *managing* (monitoring) the treatment of men with prostate cancer and not as an early diagnostic or *screening* test.[13] Clearly there is a remarkably different market share for a test confined to men with disease *versus* a screening tool for use in the routine screening of all men over fifty, and it is this distinction that Ablin complains was purposefully obfuscated for profit.[14]

The NCPC first filed for a patent on an enzyme immunoassay, or PSA test, following their publication of a 1980 paper in which they argued that PSA levels correlated to the extent of malignancy in the prostate of cancer patients *and* could detect early pathological changes in otherwise asymptomatic patients.[15] In spite of the major advances in understanding the basic science of prostate cancer thanks to the introduction of hormone and other therapies around the mid-century, efforts to control or curb the disease clinically remained frustratingly elusive. As Murphy himself said in a summary piece for the journal *Ca—A Cancer Journal for Physicians*, published in 1974: 'Despite numerous clinical advances and innovations with hormonal palliation, age-adjusted death rates for prostatic cancer have not significantly changed in the past 40 years.'[16] The possibility that prostate cancer could be detected early, perhaps while still encapsulated in the gland and so perhaps susceptible to curative surgery, was, therefore, a dazzling prospect.

In 1987 a team from Stanford University led by Thomas Stamey published a landmark study of PSA in the serum samples from 699 patients, 378 of whom had prostatic cancer.[17] Here, Stamey compared PSA to prostatic acid phosphatase (PAP) to see if the former gave more insight into malignancy than the latter, concluding that the 'concentration of serum PSA is proportional to the clinical stage of prostatic cancer in untreated patients; it is 5 to 16 times higher than that of PAP. More important, PSA is also proportional to the volume of cancerous tumour within the prostate'.[18] In underscoring that PSA appeared to be a better tumour marker than PAP, this study and another published two years later, pointed to the conclusion that PSA levels might have a positive correlation with increased tumour volume.[19] Stamey concluded that serum PSA level *might* become a useful tool for cancer *detection*, as well as a means to measure the responsiveness of a tumour to anticancer therapy, and to monitor for the recurrence of a cancer following treatment. Others went much further.

In 1991 a landmark paper in the *New England Journal of Medicine* announced the work of a Washington School of Medicine group of researchers led by the urologist, William Catalona, in developing PSA as a screening tool: '[Our] results indicate that serum PSA measurement is a useful adjunct to rectal examination and ultrasonography in detecting prostate cancer. Although all three have the ability to predict cancer, the predictive value of serum PSA levels was greatest.'[20] Catalona's team further concluded that while PSA was by itself an imperfect screening tool it could serve as a useful first warning: 'PSA measurement identifies patients at high risk; at the discretion of their physicians they may be either followed with repeated rectal examinations and PSA measurements or evaluated further with ultrasonography, biopsy, or both.'[21] Catalona would go on to be one of the most visible champions of the PSA test, promoting the cause of screening as vigorously as he defended the practice from its critics.

Taken together, Ablin sees publications by Murphy, Stamey, and Catalona as the turning point in the transformation of the PSA test from niche disease management to mass screening tool. Indeed, a search of PUBMED shows that for the 1970s and most of the 1980s a few dozen articles (at most) were published year on year referencing PSA. In 1987 the number of articles mentioning PSA stood at 93, rising to 171 in 1990, 669 in 1995, and reaching a peak of discussion in 2013 with 1872 publications. The age of PSA screening for prostate cancer was at hand. Before I go into that discussion, however, I would first like to discuss a little of what prostate cancer 'detection' meant and looked like in the decades before PSA testing.

THE DETECTION OF PROSTATE CANCER FROM THE 1900S TO THE 1990S

Earlier chapters of this book have included descriptions of how prostate cancer was 'detected' at different time periods—from observed signs and symptoms (and some physical examination) of the patient in the early period, through to use of the microscope by the turn of the nineteenth century, all the way through to quantified chemical analysis, x-rays and other kinds of visualization by the mid-twentieth century. These accounts shared something in common in that they typically arose from the diagnosis of some kind of *clinically* apparent disease, and/or were observed as part of some gross pathological anomaly at autopsy. As I mentioned in Chap. 3, anecdotal reports of *non*-clinically apparent carcinomas did exist, often buried in descriptions of surgeries undertaken to relieve some supposedly benign conditions, in the surgical literature of the nineteenth century, and it appears that many urological surgeons accepted that 'latent' cases of prostate cancer did exist. With the rise of the cellular understanding of cancer in the late nineteenth century (Chap. 2) academic pathologists—who, like physiologists, underwent a rapid professionalization phase around that time, forging careers largely disentangled from the obligation to practice medicine[22]—brought their refined skills with the microscope to bear on twentieth century debates about the natural history of cancer. One effect of this was that debates in urological surgery over the hopelessness or otherwise of surgery in cases of prostate cancer (Chap. 3) began to engage more with the question of what might be if asymptomatic cancer ('latent' or otherwise) could be better understood and perhaps subject to surgery.[23]

As I have already alluded to in Chap. 3, the consensus view emerging from debates like these was that early 'enough' detection was an elusive and perhaps pointless goal given the apparently incongruent and impenetrable behavior of prostatic carcinomas; surgical intervention was not generally favored due to its inherent dangers and apparent futility. It is important, nevertheless, to recognize that questions about the histology and natural history of prostate cancer were actively pursued throughout the period between the work of Hugh Young and Charles Huggins. This was a period of pessimism but not of silence on matters of cancer therapy.[24] This work also helps to explain and situate a piece of research work that would have an immediate and lasting impact: Arnold Rice Rich's cadaver studies published in 1935.

Rich was a pathologist at Johns Hopkins who was able to take advantage of a special feature of his department and something that was still relatively unusual even in academic medicine, that is, the routine collection of full sets of organ and tissue samples from cadavers brought in for autopsy. The custom meant that Rich had access not only to an extensive pool of pathological specimens but also of course to a large number of preparations of normal tissue (or apparently so in gross examination). Rich explained how he had decided to look at 'normal' prostate tissue in order to better understand a question that had long intrigued him:

> For a number of years the writer has been impressed by the frequency with which small carcinomata have been found in the prostate in the routine autopsy material of this Department. It seemed that these small tumours, which had attracted no attention clinically and which were brought to light unexpectedly at autopsy, were being encountered much more often than the usual estimates of the frequency of occurrence of prostatic carcinoma would have led one to expect.[25]

In following his curiosity, Rich had helped further move the nature of the debate, from centuries of anecdote and observational accident concerning the prostate to a (rough-and-ready, early) version of an organized cross-sectional prostatic study. In doing so he provided another voice—in his case a particularly compelling one—for the view of prostate malignancy as a common, but not necessarily deadly cancer (this view, that many men died *with* rather than *of* prostate cancer, would be comprehensively confirmed in later decades). In the context of the earlier debates described above, Rich's work attracted considerable scholarly attention and inspired similar studies both in the US and in Europe.[26]

As might be expected, not all investigators accepted the pessimism of the age, and the historian Robert Aronowitz has described how the most concerted and high profile research effort into cancer detection around the mid-twentieth century faired in its attempts to instigate transformative change in the field of urology.[27] In the 1950s, the Johns Hopkins trained New York urologist Perry Hudson started what would become the 'Bowery series' helped by a team of researchers from Columbia University. Aronowitz argues that Hudson aimed to build on Rich's research in two important ways: first, by providing further evidence about the general incidence of prostate cancer in men of different ages; and second, by showing how additional histologic characterization of prostate carcinoma might better inform (and encourage) active treatment in cases of 'early stage cancer'.[28] Unlike Rich, Hudson's vision was not concerned with specimen slides prepared from cadavers so much as it was about observing and sampling the prostate in the living 'patient' (the men of the Bowery series). His technique was to open up the prostates of his study cohort—a group of more than twenty-one hundred alcoholic and destitute men, drawn in to the study by the promise of 'total care' at the Delafield Hospital study site —via perineal incision followed by a biopsy. If the biopsy specimens showed signs of cancer then the Bowery men typically underwent radical prostatectomy and surgical castration (orchiectomy), followed by the new anti-androgen hormone treatment with diethylstilbestrol (DES). While the study did confirm a high incidence of unsuspected cancers in 'normal' men (just as Rich and several others had suggested), it broke little new ground in terms of diagnosis or therapy. The physical work-ups that Bowery men were subject to were standard practice in cancer care: digital rectal examinations, cystoscopy, x-ray pyelography (observing the passage of radiolabelled dyes through the urinary system), and blood tests for phosphatases, but already widely recognized as being far too unwieldy for mass screening efforts of healthy men. The routine application of the highly invasive procedure of perineal biopsy was similarly beyond practical imagining, and, as Aronowitz notes, this part of the study attracted the sternest criticism from his peers, further denigrating Hudson's methodology and reputation. As Aronowitz says, the Bowery series 'was a situation made possible by the compliant clinical material', or, to put it another way, of by the presence of a group of largely uninformed and desperate men.[29]

The biopsy techniques used by Hudson and others were little changed or improved from those of Hugh Young and the other early twentieth century urologists described in Chap. 2. This is not for lack of trying, but attempts at technical refinement were bedeviled by inconsistent sampling results. In 1922, for instance, a New York urologist, Benjamin Barringer, published the data from his own studies into a perineal punch biopsy tool noting that while various techniques of needle biopsy were by then in common use amongst specialists, no technique (including his) was free from misleading inaccuracies.[30] While positive biopsy results were meaningful, surgeons widely recognized that negative results did not rule out cancer— a likely factor in Hudson's decision to use 'open' perineal biopsies. As Cornell University urologist, Harry Grabstald, commented in his 1965 review of available techniques, 'Physicians with more extensive experience in perineal punch biopsy are strong advocates of the procedure. … One may conclude that the accuracy of the procedure depends to a great extent upon the experience of the examiner.'[31] Much like the prostatectomies so vaunted by Young in the early decades of the twentieth century, it seems that the elite knowledge, resources, and support enjoyed by the leading academic clinicians continued to be out of the reach of ordinary, rank-and-file practitioners who were likely still sceptical that they themselves could hope to replicate similar successes with their own patients.

Citing Hudson's 1955 paper on the value of transurethral biopsy (that is, biopsy taken through a puncture into the prostate via the urethra) to detect early carcinoma,[32] Grabstald commented that the technique was widely regarded as having limited use since most tumours first arose in the posterior of the gland.[33] Another transurethral surgery, though, the transurethral resection, a nineteenth century technique, was beginning to experience something of a revival in the 1960s.[34] The practice was used in a variety of urological surgeries from intervention in bladder tumours to the relief of prostatic hypertrophy by shaving the obstructing tissue, and it was the use of this latter procedure—the transurethral resection of the prostate (TURP)—that began, once again, to uncover unexpected malignancies in supposedly non-cancerous glands.

In 1990 a wide-ranging retrospective study of Medicare data from the 1970s and 1980s examined TURP with the benefit of some hindsight. The review was headed by the NCI epidemiologist, Arnold Potosky, who, together a multidisciplinary team of health system experts, collated health records data to gain a clearer understanding of the relationship between TURP use (as determined by a systematic review of patient discharge records) and changes to the reported prevalence and incidence of prostate cancer ('prevalence' being a term signifying the total number of cases present in a population at any one time; 'incidence' by contrast being a measure of *rate* of change, of *new* cases of appearing over a specified time—a month, a year, and decade and so on):

> The results demonstrate a strong correlation between the recent increase in the reported incidence rate of prostatic cancer, especially the rate of cancers classified as locally contained to the prostate, and the increasing TURP discharge rate. ... Evidence from studies of the prevalence of incidental cancer discovered upon examination of prostatic tissue after prostatectomy for treatment of clinical benign prostatic hypertrophy is consistent with the strong association between the reported increase in the incidence rate of localized cancer and TURP.[35]

While Potosky's group did determine a strong correlation between the technological intervention and a rise in prostate cancer incidence, they were keen to note that not all of the increase in *prevalence* was likely attributable to these unintended consequence of TURP. Quite deliberate (and intended) work on detection of metastatic disease via new methods of bone analysis devised during the 1970s and 1980s had created opportunities for more men with recurrent or late-stage cancers to be better monitored and treated, likely resulting in more men living longer with their disease (so adding to the total 'pool' or prevalence of cancer cases).

In addition to incidence and prevalence measures, the health system researchers also looked for changes in mortality rates over the preceding five decades and here they uncovered something that surprised them: a dramatic overall increase in deaths from prostate cancer, particularly amongst non-whites[36] (The apparently much poorer clinical outcomes seen particularly in African American men as compared to white men had been mooted since at least the 1970s, and I will return to this issue in in Chap. 8.) Potosky and his colleagues believed that the overall increase in prostate cancer deaths very possibly represented a real increase in disease risk:

> In conclusion, this analysis suggests that the recently observed increase in prostatic cancer morbidity has been primarily due to increased detection of tumors that formerly went undiagnosed. The increased use of TURP to treat benign prostatic hypertrophy is the major reason for this trend. However, the poorly understood natural history of tumors detected as a result of TURP makes it difficult to conclude that there has been no change in the risk of prostate cancer. Finally, certain patterns in long-term mortality rates, particularly among nonwhites, also suggest that the risk of prostatic cancer may have been increasing.[37]

What such risks were and might be are beyond the scope of this book, but suffice it to say that concepts of risk unearthed by Potosky's team, like many of their other findings, remain important questions to the present day.

The use of TURP decreased sharply in the late 1980s as highly effective drug treatments for benign prostatic hypertrophy came onto the market so reducing the need for mechanical intervention. As brief as this episode was, it provides a good mixture of problem 'strands' in the vexed history of prostate cancer detection that we might usually follow both forward and backward in time. Taken together, the material tools and research methods discussed in this section serve as a good reminder that medical technologies can and do radically alter and shape our ideas of the normal and the pathological. At a causal glance, technologies seem to offer a straightforward means of penetrating, controlling, or at least rationalizing disorder. A more sustained and serious look might show us a flip-side, one in which technologies themselves become powerful agents of disorder as users attempt to wrangle the natural world: multiplying, not reducing uncertainty; creating new wrinkles in the search for objectivity. Of course, it might well be in the course of time that the sciences will be perfected and all uncertainties will resolve, but that is hardly the situation in which we now live. In this reading of technology and medicine, we might expect a disease like prostate cancer, in which complexities and stubborn unknowns abound, that the knowledge produced by interactions between (ever more) powerful tools of detection and the clinical and 'sub-clinical' populations subject to them would be unlikely to result in any kind of straightforward knowledge. Whatever the veracity of my technological argument, it was certainly the case that not only was the next phase in knowledge creation around prostate cancer and detection technology *not* straightforward, it was unprecedented in the scale of the acrimony generated between the different sides of the debates.

The new technologies of detection arriving in the late 1980s and early 1990s included such devices as the spring-loaded biopsy 'gun', and the miniaturization of ultrasound devices that made transrectal ultrasound (TRUS) possible (which in turn encouraged even greater confidence in biopsy procedures as the ultrasound allowed greater visualization of the target site), a combination that saw biopsy rates amongst US men began to soar in this period.[38] One plausible explanation for this is that the combination of TRUS and the biopsy gun not only improved clinician confidence, it also made investigations of the prostate (now easier and quicker than ever) a procedure that could be reliably carried out in an outpatient office. The significance of a technology 'traveling' in this way is hard to overstate. Making tools that did not require high levels of support in terms of staff, physical plant, and other resources, created entirely new ways in which those tools (and the information they produced about the body) could circulate in and interact with the world. The unfettering of detection technology from its previous high-maintenance setting meant that costs went down for all parties, so going some way to explaining the biopsy spike referred to above (while cost and ease of use *were* important in the upward trend of biopsy use, as I will describe below I do not think that they were the only important factors). One obvious and immediate consequence (or at least obvious to anyone paying attention to their recent history) of more biopsies was a sharp rise in the incidence of prostate cancer, as more detection led to more diagnosis. Indeed, NCI data show an astonishing more than forty per cent increase in prostate cancer incidence in just the three years between 1989 and 1991.[39]

A 1995 paper authored by some of the same NCI researchers involved in Potosky's original 1990 study reported on similarities between biopsy and TURP, bluntly posing the question of whether 'increased medical surveillance and more aggressive detection have fueled the recent surge in incidence rates'.[40] The major difference between the two technologically-driven phenomena was of course the intent behind the implementation of the different tools: '[i]ncidental detection' argued Potosky and his colleagues, 'has given way to intentional detection'.[41] While noting that the FDA had approved the serum PSA test for post-treatment monitoring of known cancer patients, and for the detection of prostatic malignancy in men over fifty in conjunction with an abnormal digital rectal examination, the authors stressed that it was not, in and of itself approved as a *screening* tool, and yet, they said: 'Approximately half of the needle biopsies performed on men older than 65 years [they were again using data from Medicare] were preceded within sixty days by a PSA test. Whatever the original intentions of the FDA, a high reading on a PSA test had very rapidly become associated with a high likelihood subsequent biopsy once the test entered the marketplace.'[42] This was a 'surprising' finding according to the researchers:

> Detection of prostate cancer has increased regardless of age, race, and geographic region. The rapid increase in the use of PSA screening at all ages is somewhat surprising since several observational studies have demonstrated

that life expectancy without treatment is nearly identical to survival following definitive therapies in men aged 70 years and older. Thus, the increasing detection of prostate cancer in older men may incur a considerable psychological and physical burden without the benefit of extending years of life.

Physician uncertainty and disagreement about the effectiveness of various interventions have been associated with wide geographic variation in the use of different medical procedures. The wide variation between Seattle and Connecticut in the use of PSA screening underscores the uncertainty about the effectiveness of early detection and varying preferences for screening. The more frequent use of PSA screening in Seattle compared with Connecticut is almost certainly related to higher reported incidence rates in Seattle vs Connecticut, although the gap is narrowing with time. ... There also appears to be a relationship between early detection of prostate cancer, incidence, and radical prostatectomy, since Seattle has one of the highest rates of prostatectomy and Connecticut has the lowest.[43]

What the authors point to but do not say explicitly was the effect of 'bundling' in detection technology, whereby each part of the detection system evolved to be mutually defining, and ultimately mutually vindicating. While doctors and patients were hardly passive bystanders to these developments, clearly (as evidenced by the biopsy spike) many doctors and many patients were comfortable to move seamlessly between decisions about screening to decisions about biopsy without stopping to look for an off ramp.

The Rise of Biopsy and the Overdiagnosis Critique

By the time Potosky's updated review on detection technology and cancer incidence was published in 1995 the cancer surge had peaked and rates of incidence were on the decline. The Dartmouth Medical School academics Gilbert Welch, Lisa Schwartz and Steven Woloshin interpret that data as follows: 'the rate did decline, as the reservoir of prostate cancer available to be found dried up and as more doctors became concerned about overdiagnosis, particularly in elderly men. *But it never returned to the level it was at prior to the introduction of PSA screening* [my emphasis].'[44] They date the beginning of the decline to around 1992 (coincidentally the year that the American Cancer Society endorsed PSA screening), because, as they say, the presence of PSA screening in American society had simply become so ubiquitous that there simply wasn't a meaningful unplumbed pool of new patients to draw from. The second point that the Dartmouth team raise concerns the concept of 'overdiagnosis', a critique that in some ways harks back to the reticence of surgeons a hundred years before to (over)intervene in a sick body lest they damage or destroy what vitality their patient might have left. The late century version of this appraisal of the merits of intervention was nevertheless rather different in at least one important respect: in this articulation, overdiagnosis could mean further damaging a sickly patient, yes, but new circumstances also raised the very real possibly that severe damage could be done to a healthy patient who might, without the intervention, have experienced little or no risk from the disease itself.

Other old questions and concerns continued to linger in the age of high technology. Of most relevance here are the questions of whether latent forms of prostate cancer *did* really exist? *Could* some cancers, as some argued, be simply left alone in the patient without significant concern that the carcinoma would grow to an extent that would seriously threaten his health or his life? A 1996 study examining the prostates of one thousand men autopsied after accidental death echoed many of the findings of Rich some six decades earlier in showing once again that substantial amounts of undetected cancer could be reliably expected in men who had been ostensibly 'healthy' prior to their fatal trauma. Even men in their twenties, the study found, showed a prevalence of prostate malignancy as high as one in ten, a figure that rose to more than four in five for men who had been in their seventies when they died.[45] What might we conclude from this? How might the 1990s clinician have absorbed the decades old idea that the prostate could be cancerous but not in and of itself deadly? The simple answer to this seems to be, not all that well. Such a crude statement would, though, do an injustice to the doubtless many well-intentioned clinicians who opted for the highly beguiling 'better safe than sorry approach'. For Welch and other critics of overdiagnosis, however, there *was* no obvious safe approach, and the fact that the rhetoric surrounding screening technologies often suggested otherwise (as I discuss below) was of enormous and urgent concern. Such was Welch's antipathy towards this lionized and simplified view of screening that he, along with his Dartmouth colleagues, publicly criticized the US Postal Service after it issued a stamp promoting 'prostate cancer awareness' through 'annual checkups and tests', something, they said that, endorsed, advertised, or otherwise 'put a stamp of approval' on unproven and possibly harmful practice.[46]

Using Surveillance, Epidemiology and End Results (SEER) program data from the NCI, Welch, Schwartz and Woloshin estimated that in the first decade of the new millennium there was a sixteen percent chance that the American male would be diagnosed with prostate cancer in his lifetime but only around a three percent chance that he would die the disease.[47] While improving treatments played their role in sustaining this relatively low rate of death, the authors stressed that the data strongly suggested that many men with non-life-threatening cancers (and perhaps even more men with a 'suspected' cancer that turned out to be nothing) were subjected to unnecessary and harmful interventions:

> Without doubt, all of these men have been made to suffer from the anxiety associated with a cancer diagnosis. But the bigger issue is all the extra treatment. Most have been treated with surgery or radiation. Surgery for prostate cancer (radical prostatectomy) has known harms: roughly 50 per cent of men experience sexual dysfunction; a third have problems urinating; and a few, one or two out of a thousand, die in the hospital following surgery. Radiation can also lead to impotence and urinary problems (although somewhat less frequently), and it has a unique harm: radiation can damage the organ that sits immediately behind the prostate—the rectum. About 15 percent of men treated with radiation develop a 'moderate or big problem' with defecation, generally pain or urgency. While they cannot benefit at all, overdiagnosed patients can be grievously harmed by cancer treatments. It's not a small problem—over a million men have been overdiagnosed.[48]

In another example of the problems attendant on overtreatment as a result of overdiagnosis of prostate cancer, a 2011 study from the Brady Urological Institute at Johns Hopkins reviewed Medicare data from 1991 to 2007 to assess rates of post-biopsy hospitalization to order to look for possible red flags (that is, complications as a result of the intervention). The study's authors included both the most common types of records on biopsy—those coming from men sent for biopsy as the result of screening (either from an elevated PSA level or abnormal digital rectal exam, or both)—as well as those records associated with the much smaller group of men undergoing biopsy as a form of 'active surveillance' of a known tumour. Taken together, the two sets of records represented over one million procedures performed annually across the United States, with hospitalizations due to complications arising in one of every twenty-four procedures done; a figure the authors felt was troublingly high.[49] What's more, the Hopkins investigators interpreted their data as indicating that complications due to infection were actually on the *increase*:

> A likely explanation for the increase in infectious complications is increasing antimicrobial resistance. The American Urological Association recommends antimicrobial prophylaxis for all patients undergoing prostate biopsy with fluoroquinolones considered the antimicrobial choice based on randomized studies from the 1990s.
>
> Nevertheless, fluoroquinolone resistance has increased in the last decade.[50]

As prostate biopsy is done transrectally, the introduction of bacteria from the rectum into the prostate is a common occurrence so the issue of antibiotic resistant strains of infection a serious problem for the safety of patients and the sustainability of the procedure.

Might it be the case, though, that the overdiagnosis critics were exaggerating or, to put it another way, that injuries were in some sense 'worth it'? If some men were unfortunately harmed and many others received treatments they simply didn't need wasn't that justified if other men were saved from a death from prostate cancer? An answer to these questions first required an answer to what was in many ways *the* question: had the widespread use of screening and the subsequent increase in detection in the late 1980s done much to reduce the overall mortality related to prostate cancer? Early in the 1990s, two large-scale randomized clinical trials (RCTs) were launched to address exactly this question: did being screened *for* prostate cancer actually reduce the risk of dying *of* prostate cancer or not? The first of these trials to launch was the European Randomized Study of Screening for Prostate Cancer (ERSPC), a massive pan-European effort that randomly assigned 182,000 men, aged from fifty to seventy-four, to either to a group that was offered regular PSA screening, or a group that was not. When the ERSPC team reported their initial findings in 2007 they did note a significant *decrease* in mortality (of about twenty per cent) within the screened group but warned:

> The rate of overdiagnosis of prostate cancer (defined as the diagnosis in men who would not have clinical symptoms during their lifetime) has been estimated to be as high as 50 % in the screening group ... Overdiagnosis and overtreatment are probably the most important adverse effects of prostate-cancer screening and are vastly more common than in screening for breast, colorectal, or cervical cancer.[51]

In their analyses of ERSPC results, two French biostatisticians from the Institut Gustave-Roussy, Catherine Hill and Agnes Laplanche, noted that these high rates of overdiagnosis, combined with the serious medical side effects suffered by the fifty per cent or so of men treated, should be taken as overwhelming argument against the use of PSA testing in screening.[52] Similarly, in 2010 Jeannette Potts, a senior urologist at Case Western Reserve University, and her Cleveland Clinic health researcher colleague, Esteban Walker, pressed the point with regard to their own observations:

> At this large tertiary care and community medical center, PSA has performed hardly better than a coin toss in predicting prostate biopsy results, regardless of patient age. The controversy surrounding the management of low-grade prostate cancers, further magnifies the need for both scientific and ethical scrutiny of PSA and the courage to abandon it as a screening test.[53]

The US counterpart of the European trial—the Prostate, Lung, Colorectal, and Ovarian (PLCO) Cancer Screening Trial, was organized under the aegis of the NCI and ran initially from 1993 to 2001, enrolling some 76,693 men aged fifty-five to seventy-four. In the US trial, men were randomly assigned either to a 'screened' group, which offered PSA testing and digital rectal examination every year, or to a 'control' group left in the 'routine care' of their regular physicians (care that might, or might not, involve screening depending on what a participant's physician defined as 'regular care'—a feature of protocol design that came in for criticism).[54]

Unlike the ERSPC, the PLCO found no significant difference in mortality effects between screened and non-screened populations, and this was in spite of a twenty-two per cent increase in *diagnosis* of prostate cancer in the screened group. An uptick in screening and diagnosis did not then, as might be expected, appear to correlate with a downtick in death from prostate cancer. A second report from the ERSPC published in 2012 confirmed earlier observations of a reduction in deaths from prostate cancer in the screened group, but interestingly the author added the further finding that in terms of *overall* mortality (that is, death from all causes) the two screened and unscreened groups did not differ at all.[55]

Like the surgeons of the early twentieth century, the informed clinician of the early twenty-first century faced some seemingly intractable problems in trying to understand and best treat prostate cancer. One perennial question that had not changed very much across a century was how to best detect those cancers most likely to benefit from treatment; how to balance the interests of the few who *might* benefit against the interests of the many who might be exposed unnecessarily to harm. Despite the hundreds of thousands of international research hours and the millions of dollars invested in the ERSPC and PLCO trials, clear answers to these questions seemed as elusive as ever at the end of the twentieth century.

THE US PREVENTATIVE SERVICES TASK FORCE
AND THE CHALLENGE OF EVIDENCE-BASED MEDICINE

The next attempt to bring some clarity (and perhaps some closure) to the prostate cancer screening controversy came with the efforts of the Agency for Healthcare Research Quality (AHRQ, part of the US Department of Health and Human Services and the federal agency charged with evaluating best practice). The AHRQ-appointed US Preventive Services Task Force (USPSTF) considered the efficacy of PSA screening in the detection of prostate cancer and published a series of findings from an initial assessment in 2008 to a final recommendation in 2012. PSA testing was, of course, a controversial area to wade into, but the AHRQ and its task forces were no strangers to acrimony. When the mammography task force reported on its assessment of the efficacy of mammography as a screening tool for breast cancer in 2009, the announcement was met with howls of outrage from clinicians and patient advocacy groups alike.[56] Hostile opposition was very much expected by the prostate task force as they prepared to release their final recommendation to award PSA testing the grade of 'D', meaning that the test was actively *dis*couraged for use as a *screening* tool (by contrast, the 2009 mammography task force that had attracted so much backlash had not been nearly so harsh in its assessment—mammography was awarded the grade of 'C', meaning 'no recommendation' for yearly mammograms for women in their forties).[57] The strength of the evidence against PSA screening tool came down, once again, from the actual and anticipated harms arising from overdiagnosis (harms that tended to flow from the cascade of interventions put in motion with the detection of an elevated PSA level).[58] So concerned were the task force by the likely response to their negative recommendation that they engaged an outside body in the form of Roger Chou from the Oregon Health Science University to review their data and findings. Chou and his team duly published their independent review—a glowing endorsement of the task force findings—in the December 2011 issue of *Annals of Internal Medicine*.[59]

The comeback was immediate and fierce and Chou was as much as target of it as the original task force members.[60]

The American Urological Association (AUA) was swift in releasing its own statement declaring that the USPSTF recommendations would 'ultimately do more harm than good', if they dissuaded a single man from attending his regular screening checkup.[61] A similar response article by the prominent early proponent of PSA screening, Catalona, brought together a group of similarly minded colleagues to express their opposition and dismay:

> The U.S. Preventive Services Task Force (USPSTF), a panel that does not include urologists or cancer specialists, has just recommended against prostate-specific antigen (PSA)—based screening for prostate cancer, stating that 'screening may benefit a small number of men but will result in harm to many others.' Recognizing that prostate cancer remains the second-leading cause of cancer deaths in men, we, an ad hoc group that includes nationally recognized experts in the surgical and radiologic treatment of prostate cancer, oncologists, preventative medicine specialists, and primary care physicians, believe that the USPSTF has underestimated the benefits and overestimated the harms of prostate cancer screening.[62]

The classification of USPSTF members as 'non-experts' was also a feature of the critical response to mammography recommendations. Indeed, members of the task force typically *were* (and are) primary care practitioners (aided by epidemiologists), not specialists. This choice of membership is meant to reflect the needs of the intended audience, namely a community of interested generalists and primary care practitioners in need of robust but clear tools to aid in shared decision-making with their patients.

The kinds of 'evidence' considered by the task forces was, *pace* Catalona, never intended to be drawn from individual expertise or personal experience but was, rather, an exercise in disinterested statistical, meta-analysis of data collected from clinical trials and other admissible clinical studies. This kind of 'evidence-based medicine' had its conceptual roots in the original RCT groups of the 1940s and 1950s but the emergence of a formal movement was something much more recent.[63] The Evidence-Based Medicine Working Group formed at McMaster's University, Canada in the early 1990s, was an early example of a collaborative, evidence-based practice research group in North America. For the members of the working group, evidence-based medicine had a very real potential to transform the basis of medical education and train clinicians to practice a new, more rational medicine:

> A new paradigm for medical practice is emerging. Evidence-based medi-
> cine de-emphasizes intuition, unsystematic clinical experience, and patho-
> physiologic rationale as sufficient grounds for clinical decision making and
> stresses the examination of evidence from clinical research. Evidence-based
> medicine requires new skills of the physician, including efficient literature
> searching and the application of formal rules of evidence evaluating the
> clinical literature.[64]

Meta-analysis underpinned this new framework of evidence-based medicine (EBM). As the clinician-statistician John Bailar (of Gleason grading
fame, see Chap. 5) stated in a 1997 reflection on the state of EBM, the
core principle of meta-analysis was sound: it required a close collaboration between clinician and statistician in abstracting critical information
from research reports published and unpublished—pulling and winnow-
ing from databases, bibliographies, and disregarded or otherwise abandoned studies for analysable data—all admirable aims.[65] To do that, and
more importantly, to do that well was extremely difficult, however, and
opponents of EBM seized upon the (numerous) examples of poor practice. As Bailar himself acknowledged: 'It is not uncommon to find that
two or more meta-analyses done at about the same time by different
investigators with the same access to the literature reach incomparable
conclusions. Such disagreement argues powerfully against any notion that
meta-analysis offers any assured way to distill the "truth" from a collection
of researcher reports.'[66]

Another prominent advocate of meta-analysis and one of the UK architects of the EBM-movement, David Sackett, reflected on the critiques of
EBM as a reductive, 'cookbook' type approach to clinical care, one that
ignored the skill of the physician and the idiosyncrasies of the individual
patient:

> Good doctors use both individual expertise and the best available external
> evidence, and neither alone is enough. Without clinical expertise, practice
> risks becoming tyrannised by evidence, for even excellent external evidence
> may be inapplicable or inappropriate for an individual patient. Without cur-
> rent best evidence, practice risks becoming rapidly out of date, to the detri-
> ment of patients.[67]

Sackett and other EBM advocates have argued since the early 1990s that objections to the technique have largely rested on 'straw man' arguments that presupposed an unbreachable gulf between armchair judgment and real life practice. Such arguments were, clearly, still much on display during the USPSTF screening debates.

In addition to speaking to questions of expertise (which included considerable disputation over the finer points of the ERSPC and PLCO trials, both of which figured heavily in the USPSTF meta-analysis), the Catalona response to the USPSTF branched out into arguments over access and potential rationing:

> The recommendations of the USPSTF carry considerable weight with Medicare and other third-party insurers and could affect the health and lives of men at high risk for life-threatening disease. We believe that eliminating reimbursement for PSA testing would take us back to an era when prostate cancer was often discovered at advanced and incurable stages. At this point, we suggest that physicians review the evidence, follow the continuing dialogue closely, and individualize prostate cancer screening decisions on the basis of informed patient preferences.[68]

Anger and confusion was to be expected, perhaps, in a discourse so shaped by legislation both affectionately and derisively known as 'Obamacare' (or more formally, the Patient Protection and Affordable Care Act or ACA passed in 2010). In addition to some of the more publicly contested aspects of the new law—expanding the services that basic insurance policies were required to cover, and mandating that almost all working Americans show proof of health insurance as part of their tax returns or face fines, being two major ones—the ACA generated many other controversies, less well covered in the mainstream media. One of these less widely covered, but equally hotly contested, debates involved a feature of the new law that only preventative services endorsed with an 'A' or a 'B' rating should be automatically covered by Medicare and private insurance. While this did not preclude 'C' or even 'D' rated services being covered, the clause certainly raised issues of restricted access to services that were fuel to an already inflamed debate about the future of American medicine.

While the task force concept was, from the outset in the 1980s, explicitly set up to consider recommendations from an perspective of efficacy, charges that it was in fact more concerned with being a cost-cutting enterprise have dogged its history. Indeed, suspicions about the role of the federal government in assessing technological benefit based on cost-assessment has itself a longer history than that of the task force. The short-lived National Center for Health Care Technology (1979–1982), for instance, was established by Congress as an explicitly 'neutral' body meant to provide policy makers with expert assessments of the efficacy of new and existing healthcare technologies at time of intense anxiety over spiralling healthcare costs.[69] With the NIH more likely to fund trials and investigations of drugs over hardware, advocates of the Center hoped it would fill a research gap in American medicine at a time when new technologies, unprecedented in their expense and sophistication—MRIs and PET scanners, to name a couple—were coming into routine use. Along with the research into safety and efficacy of technologies, the Center also had a role in evaluating devices for Medicare, generating data that was then used to make reimbursement decisions (and where Medicare went the private insurance market often followed).[70] Not all health care providers and manufacturers supported these activities, however, and following lobbying by the American Medical Association (AMA) and the Health Industry Manufacturers Association, the Reagan administration pulled the plug on funding the Center.[71] In a statement provided to a Congressional subcommittee established to consider the issue, a representative of the AMA argued against the reauthorization of funding:

> because the relevant clinical policy analysis and judgments are better made—and are being responsibly made—within the medical profession. Assessing risks and costs, as well as benefits, has been central to the exercise of good medical judgment for decades. The advantage the individual physician has over any national center or advisory council is that he or she is dealing with individuals in need of medical care, not hypothetical cases.[72]

Like its parent organization, the AHRQ (which took on some but by no means all of the expansive remit of the National Center for Health Care Technology) has faced its own, repeated calls for defunding.[73] Similarly, recent criticisms of task forces echo the earlier accusations made by the AMA against the Center of detached, non-experts making dangerous decisions about services in order to ration access to healthcare.

The feared reductions in access to PSA testing for men who wanted it failed to emerge in the months and years following the USPSTF recommendations, and the antagonism of the urological community also began to subside. In 2013, for instance, the American Urological Association backed off its aggressively anti-task force stance and even incorporated some a key USPSF recommendation into their revised guidelines, in advising against annual screening for low-risk men under or in their forties.[74] In an interview with *USA Today*, the chair of the AUA panel that revised the organization's best practice guidelines, the Johns Hopkins urologist Ballentine Carter, accepted that the 'public is very enthusiastic about screening, partly because of our messaging', but overall the 'idea that screening delivers benefits may have been overexaggerated'.[75] What the AUA did continue to stress though was shared decision-making between physicians and their patients. So what of patients? What were ordinary men making of these heated debates about prostate screening and prostate cancer?

GENDER AND POLITICS OF CANCER RESEARCH

In his *The Great Prostate Hoax* Ablin lays out how he believes the Roswell group and Hybridtech exploited the emerging 'screening *zeitgeist*' of the 1990s:

> The timing for mass marketing was perfect. Promising studies in breast mammography fueled a national explosion of cancer screening. The feminist movement brought discussions of breast cancer out of the closet. Advocacy groups, led by formidable women, lobbied on Capitol Hill for universal breast cancer screening. The mantra 'early detection leads to cure,' chanted by breast cancer advocates, would soon be embedded in our national consciousness. By the early 1980s men had grown restless for their own early detection tool. Prostate cancer had a visceral grip on men akin to that of breast cancer for women; it spoke directly to gender-based fear of premature mortality and struck at the core of manhood. Men would begin to form their own advocacy groups, using celebrity prostate cancer survivors as spokesmen.[76]

It was certainly not the case that the 'early detection' 'mantra' arose in the 1990s; both the practice of screening and the 'early detection' messaging of public health campaigns were more the creations of the early, not the late, twentieth century. It is the case, however, that such messages were strongly gendered. Campaigns against breast (and, indeed, cervical) cancer urged women's participation in screening efforts, in part, by depicting compliance as serving a moral equivalence to the righteous behaviour of responsible wives and mothers.[77] Gendered messaging has, in other words, long been part of medicine.

The creation of the screening *zeitgeist* that Ablin refers to above began with the confluence of two coevolving trends: a renewed interest in applying techniques of x-ray mammography to visualize the interior of the breast as a screening (as opposed to diagnostic or prognostic) tool; and the rise of what historian of medicine Barron Lerner calls the 'professionalization of activism' from the 1970s onwards.[78] The age of breast cancer politics was about to begin in earnest, especially as the deaths of high profile activists like the journalist Rose Kushner inspired the formation of highly-organized lobbying groups like the National Breast Cancer Coalition (NBCC). The Coalition's skilful use of media campaigns and a slick series of appearances before Congress helped to greatly expand the NCI's spending on breast cancer, including via the high profile creation of the National Action Plan on Breast Cancer (NAPBC) in 1993.[79] Through the NAPBC the NBCC was able to achieve its aims to not only increase funding for breast cancer research but to also cement, as a matter of public policy, reliable access to screening services.

When early forms of PSA screening came onto the scene in the 1990s, the tests were inevitably compared to mammography as a matter not only of medicine but also of politics. Men with prostate cancer could look to the success of NBCC and groups like the Susan B. Komen Foundation (originally formed in the 1980s by the public relations executive Nancy Goodman Brinker, to memorialize her sister lost to breast cancer), and wonder where 'their' lobbying groups were. This, of course, brings us full circle in the gender politics of health funding, as it was precisely the marginalization of women's health issues within federal funding and public policy decisions that had inspired the NAPBC in the first place. In the years since the passage of the 1993 plan, though, Lerner's 'professionalization' of breast cancer activism had apparently gone into overdrive as businesses like the cosmetic company Avon became involved with the movement, helping to create the phenomena of the Susan B. Komen Race for the Cure and helping to create the pink ribbon as the iconic and ubiquitous symbol of the recent war on cancer.[80] Lacking similar attention or symbolism, prostate cancer activists were left to ponder an apparent example of a deadly gender bias at the heart of preventative medicine. For men with prostate cancer, claiming an identity as an overlooked group stimulated and motivated activism efforts, just as it had for breast cancer sufferers a decade before. Indeed, women themselves might look to the recent history of the breast cancer movement and wonder if the success it has enjoyed has deprived attention and funding for other diseases like cardio-vascular disease and lung cancer that are the major killers of their sex, too. Nevertheless, the image of an insidious neglect undermining men's health became, and remains, a potent rallying point (I discuss this in more detail in Chap. 8).

In 1989 the Prostate Cancer (later, Conditions) Education Council or PCEC was formed by the University of Colorado urologist E. David Crawford, looking to launch the first Prostate Cancer Awareness Week (PCAW). A 1997 retrospective noted:

> Prostate Cancer Awareness Week (PCAW) has become the nation's largest cancer screening program. Since its inception in 1989, PCAW has attracted >3 million participants. The number of screening centers has increased from fewer than 100 in 1989 to a high of 1800 in 1992. In 1996 nearly 800 loca-tions provided free or low cost prostate carcinoma screening with digital rectal exam (DRE) and prostate specific antigen (PSA) testing to an esti-mated 250,000 men.

> This public awareness and education program was conceived and initiated by the Prostate Cancer Education Council (PCEC) which represents urology, oncology, patient advocacy, minorities, clinical, and behavioral research. The objective of PCEC is to raise public awareness of the disease and its generally late diagnosis by the recruitment of asymptomatic men for screening with DRE and the PSA test. The goal of PCAW is to ensure that a majority of the male population age ≥ 50 years are screened for prostate carcinoma.[81]

In a series of articles for the *New York Times*, science journalist Gina Kolata, investigated the surge in demand for PSA testing in the early 1990s, asking in a 1993 piece, 'in this time of national agony over healthcare costs, how did an unapproved test with potentially astronomical costs become entrenched as part of the nation's medical system?':

> Even doctors who say they are convinced that the test is saving lives say it would never be so popular today were it not for aggressive promotions. For example, proponents and opponents of the test agree that much of the public demand was elicited by the enthusiastic advertising of the Prostate Cancer Awareness Week, paid for by the makers of drugs to treat cancer and by makers of the P.S.A. test. The ubiquitous public service announcements

featuring celebrities like the baseball star Stan Musial are paid for by TAP pharmaceuticals of Deerfield, Ill., a maker of prostate cancer treatment. ...

To me, the whole issue of P.S.A. testing is, in a microcosm, what's wrong with our health care system,' said Dr. Peter Albertson, a urologist at the University of Connecticut in Farmington. 'Industry is pumping a lot of money into this and creating demand for a product.' Urologists benefit because the test has made the prostate 'the biggest money-maker for urologists, a large part of a urology business,' he said. Hospitals benefit because prostate patients fill many beds. So, he said, even though the test has limitations, 'sometimes people don't want to look too hard at a gift horse.'[82]

Even for its strong proponents, Kolata argued, the magnitude of marketing caused concerns over how the PSA test became so widely administered. She quotes the chair of urology at the University of Michigan Medical Center, Joseph Oesterling, as saying of the tests makers like Hybritech, 'They went around the country saying, "Doctor, you need to get a P.S.A on your patients. Start using it, start using it." The next thing, patients started coming in and saying, "Doctor, check my P.S.A."'[83] At the heart of this 'surge' then was patient demand; a demand doubtless related to the emergence of direct-to-consumer advertising of drugs and technologies that began in the US in the 1980s (I discuss this phenomenon in Chap. 6).

Ablin's take on PCEC and PCAW (and other organizations, launched in the 1990s) in *The Great Prostate Hoax* follows the argument of Kolata in suggesting the relationships between pharmaceutical firms, doctors, and their patients were strongly affected by aggressive advertising of PSA by vested stake holders in the early 1990s. The apparent 'no-brainer' ('better safe than sorry') relationship between screening tests and early detection led, furthermore, to a powerful self-vindication in the processes of professionalizing prostate cancer advocacy:

[S]ince the breakout of PSA testing in the late 1980s, advocacy groups have become a big business in their own right. Set up like corporations, they need heavy revenues to pay for themselves and to pay for their activities, such as celebrity golf tournaments that raise cash for PSA-screening drives. ...

An advocacy organization called ZERO: The End of Prostate Cancer, headed by the magnetic CEO Skip Lockwood, a prostate cancer survivor himself, has 55 corporate sponsors. It's a who's who of pharmaceutical heavyweights, among them Beckman Coulter and Abbott, market leaders in PSA test kits.[84]

A 2007 report of the Washington DC-based National Cancer Coalition (the forerunner of ZERO), titled 'The Prostate Cancer Gap: A Crisis in Men's Health', described what the organization saw as discrepancies in awareness, funding, media coverage, and research between prostate and breast cancer: 'Year after year, the prostate cancer community has received less attention and less funding than many other diseases', Richard Adkins the CEO of the organization told the news organization *Bloomsberg*, just one example of the 'glaring disparities' in attention to men's health.[85] The timing of the summer 2007 report and the interview was significant. In May of 2007 the FDA had refused to approve the new immunotherapy drug Provenge, a decision that led to bitter pushback and recriminations from prostate cancer advocates, some of whom accused members of the advisory panel of blatant corruption.[86] The drug did go on to be approved in 2009, but its subsequent use and history have been bogged down in controversy and accusations of bias on the part of both its advocates and detractors.[87]

The National Cancer Coalition had many admirable goals: to increase the funding that the NCI devoted to male cancers; to increase media coverage of men's health and so to increase the likelihood that men would be better informed and more willing to discuss their health concerns; and to lobby health insurance companies to include coverage for cancer screening for men who wanted it. As has been the case with breast cancer advocates and their tenacious support of universal, yearly mammography-screening for women aged forty and over, in the face of repeated findings from the USPSTF recommending against such policies, prostate cancer advocates have tended to focus on PSA screening. As patient advocates have encouraged more men to be open about their emotional and physical health in recent years, going to get 'the test' has become a symbol of personal responsibility and masculine solidarity. In the words of the academic and popular science writer, Stewart Justman, for both breast and prostate cancer advocates, success 'was measured by numbers screened, not by improvement of public understanding' (clearly an altogether much more difficult outcome to measure).[88] Prostate cancer advocates also followed breast cancer advocates in making good use of high profile public figures in this regard—quite literally in some cases: several high profile mammography advocates from the NAPBC joined the efforts to get PCAW off the ground.[89] The US hero of the first Gulf War, General 'Stormin' Norman Schwarzkopf, diagnosed and treated for prostate cancer in 1993, became the first celebrity spokesman for the PCAW and PSA screening.

Like breast cancer activists before them, prostate cancer activists were prone to hyperbole when cancer screening was criticized. In 2002 two physicians, Gavin Yamey and Michael Wilkes, who were at that time serving as consultants to the US Centers for Disease Control and Prevention (CDC), reported on their experience of push back from activists. The pair had written a piece in the *San Francisco Chronicle* on the PSA testing and subsequent prostate surgery of local baseball hero, Dusty Baker, criticizing the newspaper's lack of mention of how controversial screening for prostate cancer was, and why. The response to their piece was immediate and alarming:

> We wrote to the *Chronicle* arguing that the newspaper had failed to reflect the massive controversy surrounding prostate cancer screening. The *Chronicle*'s editorial team knew nothing about the controversy, which is no surprise given the dominance of the US media by the proscreening lobby. The editors invited us to write an opinion piece discussing the reasons why men should not be screened. ... Within hours of our piece being published, prostate cancer charities, support groups, and urologists around the country had circulated a 'Special Alert' by email. This community has huge faith in PSA tests, and it did not care for our opinion. The email, under the header 'ATTENTION MEN!!' urged the community to take action. By the end of the day, our email inboxes were jammed with accusations, abuse, and threats. We were compared to Mengele, and accused of having the future deaths of hundreds of thousands of men on our hands.[90]

The piece in the *San Francisco Chronicle* would ignite a bitter battle for Wilkes not just with patient advocates but also with his own medical school at the University of California, Davis. Wilkes later said that he was in part moved to write the *Chronicle* editorial thanks to UC, Davis sponsoring a free public seminar on men's health, advertised under the byline, 'Prostate Defense Begins at 40'.[91] Wilkes objected to the invitation of the football hero Guy McGuire to promote a message of screening, as he did to the fact that McGuire's appearance and the event itself were largely paid for by Intuitive Surgical, manufacturer of the da Vinci robot used to perform prostatectomies. Thanks to his protestations and the *Chronicle* article, Wilkes claims to have been bullied by senior personnel at his university, although this was strongly denied by the university administration itself.[92]

When the USPSTF recommended against routine PSA screening in 2011, advocacy groups again swung into action, much as breast cancer activists had three years before in protest at mammography findings. Backed by the AUA, patient groups strongly protested the recommendations and lobbied state and federal government to protect payments for PSA screening, something that produced actual legislative intervention in places—New Jersey, for instance, passed a law in 2012 opposing the Task Force's recommendations after intensive lobbying efforts organized by the New Jersey Patient Care and Access Coalition, a group organized to represent the interests of urological healthcare in the state.[93]

The concept of early detection of cancer is a powerful one, based as it is on a kind of 'intuitive' or 'commonsense' appeal. 'It speaks to a sense of individual responsibility and the opportunity to improve one's destiny through action', argues one of the authors of the PLCO trial, Barnett Kramer. [94] For individuals seeking out screening, a 'virtuous cycle' is established whereby a negative result provides apparent guarantees of health, while a 'positive' result provides a gratifying sense for physicians and their patients that the disease has been 'caught early'. Quoting the Roman playwright Terence, Kramer reminds us that, 'One easily believes what one earnestly hopes for.'[95] It does not seem likely that the appeal of screening will lose its shine any time soon, or that the controversial and counterintuitive claims of overdetection and overdiagnosis will do much to alter the behaviour of the clinicians who order the tests or the patients who ask for them. In a medical marketplace so shaped by litigation as the US, it might simply prove too risky for clinicians *not* to go ahead and order tests (in an act of 'defensive medicine') in case a patient who later develops clinical cancer then sues.[96] Practice decisions, in other words, do not occur in a vacuum.

CAUSES AND EFFECTS IN THE PROSTATE CANCER EPIDEMIC AND ITS AFTERMATH

In spite of its short duration, the Bowery series served, according to Aronowitz, as a highly significant and 'prescient' signal of what was to come in the science of cancer prevention at the turn of the twentieth century:

> Many elements of the Bowery series—screening asymptomatic people, mass biopsies, and resulting transformation of prostate cancer into a curable disease—are now in place. And this transformation has occurred without any

profound new etiological understandings or dramatically new therapeutic principles and modalities. ...

Hudson and colleagues had demonstrated a screening plus radical intervention paradigm for which fellow urologists and ordinary men were not yet ready. It is no longer unimaginable that men without symptoms will readily consent to having bits of their prostate gland taken out and examined for cancer. In the decades since the Bowery series, there has been a great deal of tinkering with different elements of that program that have made similar practices more palatable to doctors and patients. This tinkering has catalyzed changes in medical routines and created the conditions—especially a large cohort of men at high risk of prostate cancer—for rapid, self-sustaining attitudinal and behavioral change.[97]

Aronowitz argues that while we tend to have confidence in two basic types of persuasion in medicine—either a demonstration of efficacy of one intervention over another, or by an identification of underlying causes or mechanisms of disease and the application of an intervention to remove or negate them—modern prostate cancer screening is not dependent upon either.[98] It has, rather, emerged within a style of consumerist logic where, 'the selling of fear and uncertainty makes the technologies that promise to banish them irresistible'.[99] The rise of mass screening for prostate cancer came rapidly and did not wait for the organization of clinical trials to evaluate the efficacy of PSA testing. Men were often told of their abnormal reading as a result of a routine blood test and then shown a world of impressive, new interventions like robot surgery or proton therapy (discussed in Chap. 7) promoted as more effective against cancer than older treatments, and much less liable to cause the kinds of complications that left men incontinent and impotent. This pressure to act, and to act aggressively, Aronowitz argues, makes it hard to say no or to take things slowly, something he says casts doubt on whether these men (like their Bowery predecessors) really do in fact make entirely 'informed' decisions. For their part, the major urological organizations strenuously deny that a positive PSA reading, 'automatically' results in aggressive intervention, but, as discussed in Chap. 7, the marketing of such technologies are everywhere on the US cancer 'scene' and a huge part of the economy of the healthcare industry.

Ablin's *The Great Prostate Hoax* is not the only controversial *J'accuse* book to have been written about the politics and economics of prostate prevention. In 2008 Justman published his *Do No Harm: How a Magic*

Bullet for Prostate Cancer became a Medical Quandary, investigating the origins and impetus behind the promotion of the chemotherapy finasteride as a drug that ought to be widely implemented as a preventative measure against prostatic cancer. As described in Chap. 4, the fact that most prostate cancers were hormone-dependent had been well known since the discoveries of Charles Huggins in the 1940s. Laboratory studies of finasteride generated so much excitement because the drug seemed to interfere in the metabolic breakdown of testosterone into the more potentially cancer-promoting androgen, dihydrotestosterone. The main focus of *Do No Harm* is the execution and aftermath of the Prostate Cancer Prevention Trial (PCPT) coordinated by the Southwest Oncology Group or SWOG, launched in 1994 (one of the original cooperative clinical trials groups organized under the NCI, see Chap. 5). The study made use of more than two hundred study sites to randomize thousands of healthy male volunteers aged fifty-five and older to either a finasteride treatment group or to a control group administered a placebo. As a scholarly text, Justman's style is inevitably less strident than the first-person perspective that informs Ablin's plaintive narrative, but similar issues are raised: Justman suggests that issues of hubris and commercialism led to bias in the assessment of the benefits and disbenefits of finastride just as they had (by Ablin's account) in the implementation of PSA screening. Ultimately, though, Justman's account is one in which the checks-and-balances of academic medicine (or, as Justman tells it, an unwillingness to disregard the Hippocratic maxim to first 'do no harm' to the patient) ultimately win out over over-optimism, cynicism, and commercial exploitation.

SWOG's findings, The Influence of Finasteride on the Development of Prostate Cancer, published by the *New England Journal of Medicine* in 2003, reported generally favorable results but with heavy caveats. Their findings seemed to suggest that in some men the preventative treatment might actually be linked to a higher risk of developing a later, aggressive carcinoma:

Physicians can use these results to counsel men regarding the use of finasteride. It is important to stress that finasteride reduced the risk of prostate cancer [diagnosis] in a clinical trial marked by frequent monitoring for disease and was associated with an increased risk of diagnosis of high-grade prostate cancer. For a man considering using this medication, the greater absolute reduction in the risk of prostate cancer must be weighed against the smaller absolute increase in the risk of high-grade disease.[100]

Undoubtedly informed by the ongoing debate about overtreatment, SWOG also cautioned clinicians to discuss the implications of side effects:

> There is also the matter of side effects: the incidence of adverse side effects on sexual function was higher with finasteride, but the finasteride group had a lower incidence of urinary symptoms and complications than the placebo group. Using published information on the outcomes of prostate-cancer treatment, men can weigh these trade-offs in the context of their own priorities regarding the avoidance of prostate cancer as well as their urinary and sexual function to reach a personal decision regarding finasteride use.[101]

Like Aronowitz, Justman had things to say about the character and quality of informed consent: 'There can be no objection to a patient making an informed decision to take or not to take [finasteride], but if the history of PSA testing is any guide, few patients will be well informed.'[102]

As I have already discussed, part of problem was that these complex decisions were not particularly amenable to neat public health messages fitting into any kind of easy 'mantra'. They were, instead, discussions requiring both highly informed patients and highly informed consulting clinicians. This 'best-case' scenario assumes that the clinician is aware of the most up-to-date practice guidelines. While *some* lag time between guideline and changes to actual practice is to be expected, research suggests that the duration of these transitions can be measured in years rather than in weeks or months.[103] Indeed, evidence suggests that in the years following the USPSTF recommendations, new guidelines were routinely ignored or misinterpreted in doctors' offices across the country.[104]

Knowledge lags and lacuna also exist for patients, of course. In a 2009 paper published in the *Journal of the National Cancer Institute* researchers from the Max Planck Institute in Berlin tried to assess the extent to which healthy women and men undergoing mammography and PSA screening respectively were doing so on the basis of 'informed' decision-making. In their extensive questioning of screening participants across Europe and the US, they determined that 'better informed' individuals (measured as those seeking out more sources of information, including their primary care physicians) actually tended to overestimate the benefits of screening, while underestimating potential harms.[105]

Two decades after his paper that helped launch the reputation of the PSA test, Stamey and his colleagues published the results of years of follow-up studies and reflections on the phenomenon of PSA screening:

virtually all men with prostate cancer can now be detected. On the surface this would appear to be a great epidemiological accomplishment except for the disturbing fact that while prostate cancer is a ubiquitous tumor, it has an extraordinarily small death rate.[106]

In a 2004 interview Stamey described how, in his view, early confusion over the diagnostic benefits of PSA had contributed to a massive overuse of needle biopsies. While acknowledging that for individual men undergoing treatment for prostate cancer, PSA levels retained a great deal of value as a biological indicator of therapeutic response, he urged a swift shift in attitudes linking high PSA level to the requirement for biopsy:

> Any excuse you use to biopsy the prostate—and we've been using PSA as an excuse—you're very likely to find cancer. So the real need, and that's what I have PhDs and MDs in my laboratory working on all the time, is that we need to get a marker for prostate cancer that is proportional to the amount of cancer in the prostate. Then we might be able to make some intelligent decisions about who should be treated and who shouldn't.[107]

That prostate malignancy occurred in relatively high rates even in men unlikely to ever be diagnosed with prostate cancer was an issue that urologists had wrestled with from the time of Rich's work at Johns Hopkins in the 1930s, of course. While it was claimed that new techniques of biopsy introduced in the 1990s would not pick up 'clinically insignificant' cancers, later studies found this not to be the case (including SWOG's own analysis of their finasteride findings). At the time of Stamey's analysis then this source of possible overdetection persisted, but a new and related question was gaining traction: whether or not Gleason scores (discussed in Chap. 4) had been subject to a creeping 'grade inflation' during the 1990s prostate cancer epidemic, so fuelling biopsy-driven intervention.

In their 2005 article summarizing the state of the debate on prostate cancer staging and grading, a team from the University of Connecticut, Farmington noted that the apparent dramatic improvements in survival rates reported by US cancer centers required a closer look:

> Unfortunately, several statistical artifacts may be producing a false sense of therapeutic accomplishment. Stage migration and grade shift had particularly profound impacts on prostate cancer outcomes assessment. PSA testing has produced a dramatic stage migration. Contemporary patients in the United States rarely present with advanced disease. Consequently,

contemporary survival analysis include a lead time associated with earlier diagnosis that has been estimated to be between 5 and 10 years when results are compared with historical series. Epidemiologists have described this phenomenon as 'zero-time shift' or 'lead-time bias'. Patients appear to have an extension of their survival after cancer diagnosis when they may in fact have experienced no prolongation of their lives.[108]

In addition to this PSA-driven, but artefactual, increase in survival times due to shifting the diagnosis of cancer back along the timeline of the natural history of disease, the authors also argued that tumours that might been classified in a previous era as moderately aggressive were becoming routinely labelled as high-grade, leading to an apparent statistical improvements in treatment outcomes:

> [A] tumor grade shift occurred during the 1990s for men with prostate cancer. Although the Gleason scoring system itself has not changed since the mid-1980s, its application has. Several factors, including the introduction of PSA testing, transrectal ultrasonography, the spring-loaded biopsy gun, the dramatic increase in the performance of radical prostatectomy, have conspired to produce a statistically significant upgrading in biopsy Gleason scores, which has, in turn, produced a statistically significant apparent survival improvement in our study cohort.[109]

To be sure, prostate cancer was not the only cancer subject to claims like these. That changes to classification systems had led to stage migration in lung cancer and had artificially inflated survival statistics were issues of intense debate during the mid-1980s.[110] Similarly, researchers noted how changes in the classification of breast cancers in 2003 would likely show dramatic improvements in stage-specific survival, making analysis of the efficacy of new and existing treatments difficult.[111] While some researchers were sceptical as to whether stage migration and tumour inflation really mattered, others took note and turned the issue back to overdiagnosis and overtreatment.

In an editorial in the *Journal of the National Cancer Institute*, for instance, the journal's editors took up the problems of prostate cancer classification:

> We are ... concerned that grade inflation is a component of the more insidious phenomena of overdetection and overtreatment of prostate cancer. Currently, about 50 % of men in the United States have a prostate-specific

antigen (PSA) test annually, and about 75 % of men have had a PSA test. Despite a 3–4 % lifetime risk of prostate cancer death, more than 17 % of men in the United States will be diagnosed with prostate cancer during their lifetime. By contrast the lifetime risk of being diagnosed with prostate cancer in the 1970s was about 10 %. …

One large cohort study found that more than 90 % of men with organ-confined prostate cancer currently opt for treatment. With growing data that as many as five of every six men diagnosed with prostate cancer (i.e., a 3 % risk of death but a more than 17 % risk of diagnosis) may not need treatment and the evidence that treatment adversely affects quality of life, why is it that so many men opt for treatment? One reason may be our risk averse society. … Another reason, however, may be the application of outcomes of watchful waiting for prostate cancers of decades ago to a patient's tumor today with its current Gleason score.[112]

In other words, as proponents of intervention had looked to historical data to make their case, they had in fact, in the words of the lead author of the Connecticut study, Peter Albertsen, simply not compared 'apples to apples'.[113] As a result, a skewed, over determined case for intervention in preference to surveillance had emerged within US medicine. Looking more widely at how primary care physicians interpreted data about screening tests, a group of researchers from the Dartmouth Institute for Health Policy and a team from the Berlin Max Planck Institute conducted an assessment of how well clinicians understood survival statistics.[114] They found that the statistical context of screening data, especially lead-time bias, was poorly understood by the study cohort of more than four hundred inpatient and outpatient physicians who responded to their survey. The majority of participants, when given data linking increase in survival times to increases in detection rates, drew the overly simplistic conclusion that 'screening saved lives', making appeals to the problems of overdetection and overdiagnosis difficult to communicate to an audience inclined to inflate the benefits of screening.

Reflecting on the problem of resistance to changes in practice based on evidence, the urologist Antony Horan claimed that the phenomenon was not necessarily or particularly a problem of specialist urologists, but rather of generalist physicians. It was that group, he said, for whom squeamishness over the performance of digital rectal examinations was highest and critical reflection on the limits of PSA screening the lowest. Referring to Stamey's later reflections on the PSA test that he had such an instrumental role in developing, Horan said:

These statements might have been King Canute railing against the waves. Physicians other than urologists proved to be a good market for a blood test for prostate cancer. Nobody likes to do rectal exams. Some internists bought machines to do the test for profit in their offices. The number of PSA tests ordered for screening purposes by nonurologists, doubled between 1991 and 1996 in the Medicare population.[115]

Horan notes that his perspective was shaped by a career in the Veterans Administration, where a system of 'capitation' (so much money provided for so many number of patients, healthy or not) forced him to examine the literature more closely than most in order to bring some justice to the decision not to biopsy all cases of men with elevated PSA. Ironically, then, it was financial conservatism that helped Horan to better understand the importance of treatment decisions, in a healthcare economy otherwise geared towards more 'work' (tests, interventions) equaling more money.

For a man to decide *not* to get his PSA tested is to live with the risk that a prostate cancer might kill him, but it is also a decision that will save him and many, many more men who likely didn't need it, from potentially mutilating treatment. Regardless of cost-benefit analysis, it *is* a rational choice for an informed man *not* to get screened for PSA. Within a culture that determinedly marches on with an almost entirely martial view of cancer as a war against nature, a campaign to seek out and destroy alien cells, such choices seem dangerous, irresponsible even. As Justman put it in a later book, *The Nocebo Effect: Overdiagnosis and its Costs*: 'Medicine proceeded with PSA testing not only in the absence of evidence of its value but despite being aware of its traps, as if the urgency of the war on cancer overrode medical restraint, just as the terror of cancer aversion overrode the aversion men would normally feel toward harmful, especially sexually harmful, treatments.'[116] Even while the biological understanding of cancers has shown that tumours are unpredictable in their cause and effect, we as a public are still caught up in the reassuring promise of the surgeon and the knife to go in and 'get it all', to cut away and destroy the enemy within.

Sadly, not all men can live with prostate cancer—there are highly aggressive forms of the disease that kill quickly regardless of screening or treatment—but many others can and do live with their condition thanks in part to treatment and in part due to this enormous variability in how tumours progress. A remarkable transformation in our ability to intervene in diseases has led, as the great twentieth century oncologist and essayist

Lewis Thomas would have it, to many 'half-way technologies'.[117] Such technologies might extend our lives, but they do not cure us. As this is a state of affairs set to continue in the twenty-first century, it would make sense to begin to reconsider our rhetoric. We *can* sometimes, or perhaps even often, live with disease. We should not overly harm ourselves by making drastic intervention the expectation and aggressive follow-up treatment the norm. In the remainder of this book, I will focus on how the growth of new technological options and new forms of healthcare business in American medicine thrived within this atmosphere of increased rates of prostate cancer detection and treatment in the 1990s and early 2000s.

NOTES

1. Andriole et al., Mortality Results from a Randomized Prostate-Cancer Screening Trial, 1311.
2. Ablin, The Great Prostate Mistake.
3. Wang et al., Purification of a Human Prostate Specific Antigen, 238, n. 10.
4. Ablin and Piana, *The Great Prostate Hoax*, 238, n. 10.
5. Soanes, Ablin, and Gonder, Remission of Metastatic Lesions Following Cryosurgery in Prostatic Cancer.
6. McCallum, *The Prostate Monologues*, 45.
7. Ablin and Piana, *The Great Prostate Hoax*, 13.
8. Ibid., 15–16.
9. Ibid., 20–21.
10. Jones, Networked Success and Failure at Hybritech, 451.
11. Ibid.
12. Ibid., 452.
13. Ablin and Piana, *The Great Prostate Hoax*, 30.
14. Ibid., 53.
15. Kuriyama et al., Quantitation of Prostate-Specific Antigen in Serum by a Sensitive Enzyme Immunoassay.
16. Murphy, Prostate Cancer, 282.
17. Stamey et al., Prostate-Specific Antigen as a Serum Marker.
18. Ibid., 914.
19. Stamey et al., Prostate Specific Antigen in the Diagnosis and Treatment of Adenocarcinoma of the Prostate, 1076.
20. Catalona et al., Measurement of Prostate-Specific Antigen, 1160.
21. Ibid., 1161.
22. Porter, *The Greatest Benefit to Mankind*, 320.
23. Gardner, Some Remarks on Prostatic Carcinoma, 406.
24. Binney, Cancer of the Prostate.

25. Rich, On the Frequency of Occurrence of Occult Carcinoma of the Prostrate, 274.
26. Andrews, Latent Carcinoma of the Prostate.
27. Aronowitz, 'Screening' for Prostate Cancer in New York's Skid Row.
28. Aronowitz, From Skid Row to Main Street.
29. Ibid., 301.
30. Barringer, Carcinoma of the Prostate.
31. Grabstald, Biopsy Techniques in the Diagnosis of Cancer of the Prostate, 134.
32. Hudson et al., Prostatic Cancer.
33. Grabstald, Biopsy Techniques of the Diagnosis of Cancer of the Prostate, 134.
34. Mitchell, Transurethral Resection.
35. Potosky et al., Rise in Prostatic Cancer Incidence, 1627.
36. Ibid.
37. Ibid.
38. Drago, The Role of New Modalities in the Early Detection and Diagnosis of Prostate Cancer.
39. Ries et al., *SEER Cancer Statistics Review, 1973–1991*.
40. Potosky et al., The Role of Increasing Detection in the Rising Incidence of Prostate Cancer, 548.
41. Ibid., 551.
42. Ibid.
43. Ibid.
44. Welch, Schwartz, and Woloshin, *Overdiagnosed*, 57.
45. Sakr et al., Age and Racial Distribution of Prostatic Intraepithelial Neoplasia.
46. Woloshin and Schwartz, The U.S. Postal Service and Cancer Screening.
47. Welch, Schwartz, and Woloshin, *Overdiagnosed*, 47–8.
48. Ibid., 58.
49. Ries et al., *SEER Cancer Statistics Review, 1973–1991*, 1833.
50. Ibid.
51. Schröder et al., Screening and Prostate-Cancer Mortality in a Randomized European Study, 1328.
52. Hill and Laplanche, Prostate Cancer.
53. Potts and Walker, Isn't It Time to Abandon Prostate Specific Antigen, 320.
54. Andriole et al., Mortality Results from a Randomized Prostate-Cancer Screening Trial.
55. Schröder et al., Prostate-Cancer Mortality at 11 Years of Follow-Up.
56. US Preventive Services Task Force, Final Update Summary: Breast Cancer.
57. Ibid.
58. Ibid.
59. Chou et al., Screening for Prostate Cancer.

60. Carlsson et al., Screening for Prostate Cancer.
61. AUA, AUA Responds to New Recommendations.
62. Catalona et al., What the U.S. Preventive Services Task Force Missed, 137.
63. For an account of the emergence and growth of the evidence-based medicine movement see Daly, *Evidence-Based Medicine and the Search for a Science of Clinical Care.*
64. Evidence-Based Medicine Working Group, Evidence-Based Health Care, 2420.
65. Bailar, Editorial, 559.
66. Ibid., 560.
67. Sackett et al., Evidence Based Medicine, 72.
68. Catalona et al., What the U.S. Preventive Services Task Force Missed, 138.
69. Perry, The Brief Life of the National Center for Health Care Technology, 1095.
70. Ibid., 1097.
71. Seymour, The National Center For Health Care Technology.
72. Perry, The Brief Life of the National Center for Health Care Technology, 1098.
73. Johnson, House Committee Passes HHS Funding Bill That Ends AHRQ.
74. AUA, Detection of Prostate Cancer.
75. Szabo, Urology Group Stops Recommending Routine PSA Test.
76. Ablin and Piana, *The Great Prostate Hoax*, 2014, 15.
77. Aronowitz, Do Not Delay.
78. Lerner, Breast Cancer Activism.
79. Ibid., 227.
80. For an account of the aggressive expansion of corporate interests into the sphere of cancer activism see King, *Pink Ribbons, Inc.*
81. DeAntoni, Eight Years of 'Prostate Cancer Awareness Week', 1845–6.
82. Kolata, How Demand Surged for Prostate Test.
83. Ibid.
84. Ablin and Piana, *The Great Prostate Hoax*, 59.
85. Arnst, A Gender Gap in Cancer.
86. Ablin and Piana, *The Great Prostate Hoax*, 152–3.
87. Ibid.
88. Justman, How Did the PSA System Arise?, 311.
89. Ibid., 310.
90. Yamey and Wilkes, The PSA Storm, 431.
91. Lenzer, Professor Was Harassed by His University.
92. Lenzer, Professor Who Criticized Prostate Screening Seminar.
93. AUA, AUA Applauds Passage of New Jersey Legislation.
94. Kramer and Croswell, Cancer Screening, 126.
95. Ibid., 135.

96. Collins et al., Medical Malpractice Implications of PSA Testing.
97. Aronowitz, From Skid Row to Main Street, 311.
98. Ibid., 312.
99. Ibid., 314.
100. Thompson et al., The Influence of Finasteride, 223.
101. Ibid., 223–4.
102. Justman, What's Wrong With Chemoprevention of Prostate Cancer?, 23.
103. Morris, Wooding, and Grant, The Answer Is 17 Years, What Is the Question.
104. Drazer, Huo, and Eggener, National Prostate Cancer Screening Rates.
105. Gigerenzer, Mata, and Frank, Public Knowledge of Benefits of Breast and Prostate Cancer Screening in Europe.
106. Stamey et al., The Prostate Specific Antigen Era, 1300.
107. Barclay, End of an Era for PSA.
108. Albertsen et al., Prostate Cancer and the Will Rogers Phenomenon, 1250–1.
109. Ibid., 1251–2.
110. Feinstein, The Will Rogers Phenomenon.
111. Woodward, Changes in the 2003 American Joint Committee on Cancer Staging.
112. Thompson, Canby-Hagino, and Lucia, Stage Migration and Grade Inflation in Prostate Cancer, 1237.
113. Drazer, Huo, and Eggener, National Prostate Cancer Screening Rates After the 2012 US Preventive Services Task Force Recommendations.
114. Wegwarth et al., Do Physicians Understand Cancer Screening Statistics?
115. Horan, *How to Avoid the Over-Diagnosis and Over-Treatment of Prostate Cancer*, 24.
116. Justman, *The Nocebo Effect*, 145–6.
117. Thomas, On the Science and Technology of Medicine.

CHAPTER 7

Radiotherapy and Evidence in an Age of High Technology

Proton and other particle therapies need to be explored as potentially more effective and less toxic [radiotherapy] *techniques. A passionate belief in the superiority of particle therapy and commercially driven acquisition and running of proton centers provide little confidence that appropriate information will become available. Objective outcome data from prospective studies is only likely to come from fully supported academic activity away from commercial influence. An uncontrolled expansion of clinical units offering as yet unproven and expensive proton therapy is unlikely to advance the field of radiation oncology or be of benefit to cancer patients.*

Michael Brada, et al., Proton Therapy in Clinical Practice: Current Clinical Evidence (2007)[1]

It is our impression that [Intensity-Modulated Radiotherapy] *was adopted because the improvements in dose distributions which it seemed to offer were compelling—a situation not unlike that of protons—and that the clinical studies came later. ... Given that IMRT is widely used, and hence affects large numbers of patients, and proton beam therapy is not and does not, it would seem much more urgent to perform RCTs for IMRT than for protons, and we cannot understand why our critics seem to believe the opposite.*

Michael Goitein and James Cox, In Reply: Proton Therapy in Clinical Practice (2007)[2]

Some remarkable discoveries were made about the fundamental nature of the universe at the turn of the nineteenth century. In 1895, the German physicist Wilhelm Conrad Röntgen, working at the University of

© The Editor(s) (if applicable) and The Author(s) 2016
H. Valier, *A History of Prostate Cancer*,
DOI 10.1057/978-1-137-56595-2_7

163

Würzburg, observed a strange fluorescence as he tinkered with cathode ray tubes, a phenomenon he would later describe as 'x-rays'. Building on this excitingly new and mysterious finding in 1899 two Polish scientists working in Paris—Marie and Pierre Curie—announced to the world the discovery of several compounds that emitted similar radiating energy, including an element they termed 'radium' (from the Latin *radius* or 'ray'). Within a few years both discoveries were taken up into medical use, particularly in the field of cancer after the Curie's published evidence demonstrating that 'radioactivity' was likely biologically destructive to tumours (as it was to all tissues, as many of the early pioneers in this work found out at the cost of their own health).[3]

Following on from these sensational observations, several urologists, including Hugh Young at Johns Hopkins (see Chap. 2), began to treat patients with radium introduced via the urethra or rectum to try to reduce tumour activity:

> At the International Medical Congress, London, 1913, in the Section of Urology, Pasteau and Degrais presented a method for the treatment of cancer of the prostate with radium. The technique consisted simply in introducing a silver tube containing radium to which was attached a long wire which was employed to introduce the radium into the catheter to the proper distance, into the urethra, where it was left in place for an appropriate length of time. ... Being impressed with the good results which Pasteau had secured in two cases by this extremely crude apparatus, I secured 102 milligrams of radium element in a glass tube and set about to construct more accurate instruments for the introduction of radium into the urethra, prostate, and rectum.[4]

And try and perfect it Young's team did, publishing their results of treatment by radium in over one hundred cases of cancer of the prostate. While the results were encouraging, they noted the immense problems caused by radium in irritating and ulcerating surrounding tissues during treatment.[5] Over at the New York Memorial Hospital, Young's fellow urologist, Benjamin Barringer, also took up the use of radium, thanks to a large gift of the stuff to the hospital in 1915.[6] Working with his physicist colleague, Gioacchino Failla, Barringer figured out a way to encapsulate radium within a thin gold tube to make pellets or 'seeds' to be implanted directly adjacent or into the prostate itself.[7] Across town at the Bellevue Hospital, urologists Edward Loughborough Keyes and Russell Ferguson tried a combination of this local irradiation with so-called 'radioorchiectomy' (irradiation of the testicles) to similarly encouraging results.[8] Early optimism soon faded, however, when follow up studies failed to replicate earlier successes, and radiation would not become a common modality in prostate cancer treatment until the post-WWII era.

Interest in 'brachytherapy' (taken from the Greek *brachy*, meaning short) was briefly renewed during the 1950s when the Iowa surgeon Rubin Flocks began experimenting with a radioactive isotope of gold.[9] While Flock's method appeared to show some improvement in five-year survival times over hormone therapy and surgery alone, interest in the use of implanted radioactive seeds was soon overshadowed by the coming of the new tool in treatment: 'external beam' radiotherapy. The first of these devices to gain traction after WWII was the so-called 'cancer bomb' machines—developed with the support of the Canadian National Research Council in the 1950s and deployed in hospitals in Saskatchewan and Ontario—that made using the nuclear reactor-manufactured isotope cobalt-60.[10] Later known as the 'gamma knife', cobalt-60 technology provided a potent source of gamma rays that could, unlike x-rays, provide a therapeutic dose deep within the body's interior without devastating the skin and tissues in between. Around the same time, a team from the Stanford Medical Center led by one of the early pioneers of 'radiation oncology', Malcolm Bagshaw, reported on another type of external beam technology, the university's linear accelerator, as a source of x-rays (the work had, in fact, been going on for years but wartime conditions forced discretion). The Stanford team believed that, as linear accelerators were capable of producing very high-energy x-rays, the technology could be developed to treat deep tumours (such as those of the pelvis) at lower exposure times so sparing skin and healthy tissues.[11] Bagshaw remarked that the 'advent of modern supervoltage technics has permitted a re-examination of the efficacy of external irradiation',[12] further concluding that his studies had 'demonstrated a new approach to the treatment of localized carcinoma of the prostate, an approach which promises survival rates comparable to those achieved by radical surgery with less hazard to the patient and less disturbance to normal function'.[13] This hope, that radiotherapy would do away with the need to surgically intervene in the prostate and avoid postsurgical hazards such as incontinence and impotency, had enormous influence. The practice of brachytherapy continued into the latter twentieth century, notably with Willet Whitmore and his team at Memorial Sloan-Kettering Cancer Center in the 1970s, as did the use of cobalt-60 machines and high-energy x-rays produced by linear accelerators.[14] The greater and more influential phenomenon—at least as far as the treatment of prostate cancer was concerned, however—was the development of fast-particle therapy via the cyclotron (and its later iteration, the synchrotron).

As I discussed in Chap. 6, activism around prostate cancer had helped drive and sustain certain clinical approaches to screening and biopsy. In the case of radiotherapy, similar forces of advertising and strong consumer interest helped build and sustain the new, expensive technology of proton beam therapy even in the absence of data as to its efficacy compared to standard, cheaper therapies. The great selling point of protons, however, was their supposed accuracy: the task of all radiotherapy is to provide maximum lethal dosage to the tumour while leaving healthy surrounding tissue unharmed, and protons, at least on paper, appeared to offer this. As the screening, biopsy, and treatment epidemic picked up speed in the US in the early 1990s ever younger men were drawn into potentially disabling treatments, so promises of the 'trifecta' of cure, continence, and potency were quite appealing (this also helped to sell the later Da Vinci robotic surgical system to the American male contemplating prostatectomy).[15]

While proponents of evidence-based medicine wrung their hands at this market success of proton therapy, the self-styled 'protoneers', in turn, pointed their fingers at a clinical trial system that they believed to be mired in ethical and practical problems potentially undermining patient choice. In many ways, then, the intertwining of the prostate cancer and proton therapy stories during the late twentieth century is also an account of how the 'gold standard' was fairing in an age of high technology and high consumerism. Before I discuss the rise of proton therapy, however, it is important to look at the (largely failed) technology of neutron therapy that came before it. All technologies need a space in which they 'make sense' to developers and users, and sometimes, as was the case for proton therapy, such intellectual and logistical space is forged by technologies that then fade away to be largely forgotten to history.

THE RISE AND FALL OF NEUTRON THERAPY AND THE RISE (AND RISE) OF MEDICAL PHYSICS

In the early 1930s the future Nobel Prize winning nuclear physicist Ernest Lawrence led a team at the Lawrence Berkley Laboratory (LBL) at the University of California to construct the world's first cyclotron. His machine was essentially a particle accelerator, and while the idea of particle accelerators had been conceived years earlier in the work of Robert Van de Graaff and others, Lawrence's genius was to produce a machine with a circular design and an alternating voltage so allowing the acceleration of particles up to enormous, unprecedented velocities. Similarly, while the existence of the subatomic protons and neutrons had been proposed a decade and a half earlier by the 'father' of nuclear physics Ernest Rutherford (leader of the first team to split the atom in 1917), the ability to produce and study these particles was extremely limited. The Berkeley cyclotron emerged into a world of great excitement and high profile research aimed at finding experimental proof of the existence and characteristics of subatomic particles, a field these self-styled 'cyclotroneers' both reflected and shaped.[16] One very significant way the invention of the cyclotron affected the field of experimental physics, for instance, was in the application of fast moving particles, specifically neutrons, to biological matter. The work on the Berkeley cyclotron achieved this in two ways: first by bombarding elements with neutrons so turning them into radioactive isotopes able to be injected or ingested into the body; and second as a means to produce a high-energy beam of neutrons capable of penetrating the tissues of the body directly.

The first kinds of biological research by the LBL team were studies using the more than a dozen radioisotopes capable of being produced in the cyclotron. John Hamilton and Robert Stone, with the assistance of Ernest Lawrence's brother John, a physician, began to look for ways to apply their research findings in the clinical setting. This early research owed much to the vastly important discovery made by Marie and Pierre Curie's daughter, Irene, and her husband Frédéric Joliot, that stable elements could be induced to become radioactive through bombardment by other types of radioactivity or by neutrons. The interests of the philanthropic Rockefeller Foundation in supporting fundamental physics and chemistry research applicable to medical and public health problems also influenced the direction this and the second type of research at Berkeley: direct patient treatment. Lawrence and his colleagues successfully turned to the organization to supplement the meager sources of funding they had assembled from other sources in Depression-era America. As early as 1935 he wrote to his friend and mentor, the Danish Nobel Prize winning physicist Niels Bohr, 'I must confess that one reason we have undertaken this biological work is that we thereby have been able to get financial support for all of the work in the laboratory. As you know, it is much easier to get funds for medical research.'[17]

In 1942, Robert Stone, along with a University of California Medical School colleague, John Larkin, published the results of more than one hundred and twenty patients treated with the cyclotron over a twenty-two month period.[18] In the climate of ethics and human subject research in the early 1940s, such experimentation on patients elicited little soul-searching, the authors flatly stating that:

> Between December 1939 and Sept.15, 1941, 128 patients were treated. … The majority of the … patients were selected from the Out-patient Department of the University of California Hospital after complete examinations, including biopsies, had been made. About 50 per cent were chosen by the physicians of the Visible Tumor Clinic. They were patients who, in the opinion of that group of doctors, could not be cured by surgical or x-ray treatment. It was felt that neutron therapy must show decided effects in advanced cancer, as represented in these patients, before its use for the treatment of small localized lesions could be justified.[19]

The cyclotron continued to exist as an experimental physics tool throughout this period, with Stone and Larkin noting that patients were seen on one of the three afternoons a week that the equipment was made available for medical purposes.[20] The issues of patient treatment in a non-clinical setting were pressing since severe skin reactions, nausea, and other toxic reactions were common. The authors make a nod to these difficulties as they conclude that of the patients treated, '14 were so sick that they probably should not have been treated'.[21]

While the overall results on multiple types of cancer at different sites were statistically disappointing, Stone and Larkin did propose that the study had laid important groundwork in dose toleration and calculation of effective therapeutic dosages, calling for more research and noting that the principle of anti-tumour activity in fast neutrons had been proven (and there were some successes: of all the cancer types studied, the prostate series was most successful in terms of survival time, but as the authors noted, this type of cancer did not always hasten death even when at an advanced stage). For much of the rest of WWII, Lawrence and the Berkeley network turned their attentions to the cyclotron as a means to determine biohazards for those working on the atomic bomb project. When Stone and others returned to their patient series it became obvious that neutron therapy resulted in discouragingly high rate of late adverse effects, which by far outweighed marginal benefits.[22] Other researchers, perhaps most notably the English physicist Louis Harold Gray (for whom the standardized unit of ionizing radiation, the gray, is named) continued to pursue radiobiological research with neutrons at the Mount Vernon Hospital in London. It would be decades before researchers risked new clinical studies of neutrons (it turned out in subsequent studies that Stone had inadvertently used much higher dosages than clinically necessary, hence the high incidence of secondary effects).[23] The enticing notion persisted, however, that the high energy of neutrons created greater damage to cancer cells and at lower doses than conventional x-ray therapy (particularly as, as Gray proved, neutrons unlike x-rays did not require oxygen to be present in the tumour—often a low oxygen environment—to be effective) and this made the medical operationalization of neutrons an enduring if frustratingly elusive goal.

In the late 1960s, the radiotherapist Mary Catterall began a clinical study at London's Hammersmith Hospital of neutron therapy, using neutrons generated by a cyclotron that was the world's first hospital-based device when it went online in 1955. Catterall's early results were encouraging enough for the UK Medical Research Council (MRC) to commission a five-year clinical trial for head-and-neck cancers. Beginning in 1977, the RCT used patient data from the Hammersmith cyclotron and a similar neutron therapy device installed in the mid-1970s in the Scottish city of Edinburgh.[24] Disappointing endpoint data, and rankling issues in protocol design that lead to disputes over comparability of the data generated from the different sites of the trial, contributed to a spirit of scepticism and, subsequently, a decline in enthusiasm for neutron projects in the UK.[25] When the next UK cyclotron service—commissioned for the Clatterbridge Hospital in the northwest of England in 1984—became operational, its use as a neutron therapy facility (in this case, for treatments of the pelvis) was short-lived. The Clatterbridge cyclotron was soon re-engineered to deliver proton therapy (mostly to treat ocular tumours) thanks in large part to encouraging news coming out of US centres for radiation oncology in the 1980s.[26] Generally speaking, radiotherapists in the US had begun to participate in coordinated trials considerably later than their UK counterparts, but when they did become involved—as they did in earnest in the 1970s and 1980s—the impact was substantial.

As described in Chap. 5, the cooperative clinical trial program of the National Cancer Institute (NCI) was created along with the Cancer Chemotherapy Service Center (CCSC) in 1955. The program had some seventeen groups organized to receive NCI funds to study the comparative anti-cancer properties of the new generation of cytotoxic drugs alone or, sometimes, in combination with existing surgical and radiological therapy. No direct, formal collaboration between clinical radiologists existed, however, until 1963 when the then NCI Director, Kenneth Endicott, encouraged a group of fifteen radiotherapists to form the Committee for Radiation Therapy Studies (CRTS) under the leadership of the head of radiology at the MD Anderson Cancer Center (MDACC), Gilbert Fletcher.[27] Hodgkin's disease and prostate cancer were the first two types of cancer to be investigated under the CRTS, but both trials reportedly suffered from poor accrual and insufficient study control.[28] Pressure on the NCI to supply the kinds of funds necessary to run trials across multiple clinical centres (to accrue patients) and set up a statistical office (to organize and analyse data collected) lead to the establishment of the Radiation Therapy Oncology Group (RTOG) in 1968.[29] During the 1970s and early 1980s, RTOG studies focused primarily on studies of dosage—particularly the practice of 'fractionation', or giving doses in discrete packages at specified times rather than as one large dose all at once—as well as on combining radiological and hormonal and chemotherapeutic interventions.

With worldwide interest in neutron therapy so high in the early 1970s, the RTOG organized multisite clinical trials (some in collaboration with the UK's MRC), beginning with the MDACC's use of the Texas A&M cyclotron at College Station, followed by the University of Washington's use of the Naval Research Laboratory Cyclotron Facility, the Great Lakes Neutron Therapy Alliance using the NASA cyclotron, and the Fermi National Accelerator Laboratory (Fermilab) collaboration with the University of Chicago.[30] Initially, clinical trials were carried out on patients with a less than ten per cent probability of survival with conventional treatment and, despite the less than ideal conditions of attempting patient care in these nonclinical settings (they were physics laboratories), the NCI received enough encouraging data to fund further trials with more patients. The resulting studies yielded a considerable amount of data about the biological effects of radiation, but the treatment equipment and conditions of patient care continued to be suboptimal. Frequent breakdowns of experimental equipment pushed well beyond the uses it was designed for disrupted treatment availabilities; moreover, dosages were difficult to control in practice and many patients were sickened by accidental overdose. In response to these problems, the NCI was persuaded to provide further grants this time to allow hospitals to build their own neutron therapy machines onsite. As interest in medical physics began to spike in the 1970s, commercial device manufacturers began to court prominent cancer centres to collaborate in the design and construction of these new cyclotrons.

In 1981 the MDACC received the first of the NCI cyclotron-equipment grants and, along with the Berkeley, California-based The Cyclotron Corporation (TCC), they built the country's first hospital-based cyclotron in Houston, Texas; a machine that became operational in 1984.[31] The next NCI awardee, the University of Washington, contracted with a different company: Scanditronix Corporation—by then responsible for building one of Europe's early neutron generators at the Gustaf Werner Institute, Uppsala in the company's home country of Sweden. A further two centres—the University of California, Los Angeles and the Fox Chase Cancer Center in Philadelphia—were also awarded NCI monies and also contracted with TCC; they did, however, construct their cyclotrons using different designs both from each other and from MDACC. This diversity in design was purposeful as it allowed the NCI to assess comparative efficacy and protected (as far as was possible) against catastrophic design failure.

While the reliable production of neutrons was a feat easily enough achieved by the cyclotrons of the late 1970s, the ability to produce a well-focused beam of neutrons capable of a rotation through the multiple angles necessary for effective targeted therapy was not. Delays and disappointments marked these years of the NCIs fast neutron program, and the bespoke nature of these machines required a high degree of manufacturer support. So when the beleaguered TCC filed for bankruptcy in 1983 the loss of support was a considerable blow to MDACC, UCLA and Fox Chase, and a year later the neutron generator in Philadelphia was decommissioned as the strains on its operation became too great.[32] On the clinical level, results of the RTOG neutron therapy trials for prostate cancer and the (joint RTOG-MRC sponsored) trials for salivary gland cancer seemed promising initially, but, as had been the case in the UK neutron RCT results, trial designs and subsequent data analysis were strongly criticized. The use of neutrons, so effective in laboratory testing for antitumour properties, and the focus of some twenty years of optimism and multi-million-dollar research and development investments began to decline for all but the most experimental treatments for the hardest to treat cancers (such as highly malignant brain tumours where chances of survival were at best slim and treatment options were sparse). One enduring legacy did emerge from this period of neutron research, however, and that was the motivation and ability to go on exploring other types of fast particles as potential anti-cancer therapies.

Charged Particles and IMRT: New Radiotherapy for Prostate Cancer

That other energized particles, particularly protons and light ions produced by cyclotrons (and the later generations of particle accelerators such as the synchrotron, so named as it produced 'synchronized' patterns of magnetic fields to achieve more acceleration, and therefore more energy, for its particles), could be turned to therapeutic uses was not new idea in the 1980s. Perhaps encouraged by his Quaker background to bring humanitarian applications from nuclear technology, the head of the Manhattan Project cyclotron division, Robert Wilson, had suggested that protons might be turned to cancer treatment as early as 1946.[33] The source of Wilson's interest was that protons, in keeping with other charged particles, demonstrated a rapid loss of energy at the end of their trajectory—the so-called 'Bragg peak' effect. Unlike photons like x-rays and gamma rays whose ionizing radiation damaged tissues on their way to and leaving tumours, charged particles could be energized to penetrate the body and then stop at a desired location. The energy 'dump' produced at this stopping, in theory at least, offered an attractive alternative to existing therapies.[34] As had been the case with neutron therapy, there were from the 1950s to the 1970s attempts to treat cancer patients in cyclotron-equipped physics laboratories, albeit with scant success.[35]

The increasing availability of smaller and faster computers by the 1980s helped to greatly improve dosimetry calculations for all kinds of radiotherapy, transforming the arduous process of radiation therapy treatment planning.[36] Similarly, the high degree of contrast resolution achieved with the emerging technologies of computer tomography (CT), magnetic resonance imaging (MRI), and positron emission tomography (PET) scanning of the 1980s offered an unprecedentedly powerful means to image tumours and so provide well-defined targets for radiation oncologists.[37] Out of these innovations came 3-D conformal radiation therapy and intensity-modulated radiation therapy (IMRT), both premised on the promise to more precisely target the tumour while leaving surrounding tissues unharmed.[38] These developments also opened up a new pathway for the development of proton therapy, and in the late 1980s a small group of researchers interested in protons came together to form a professional organization aimed at managing and implementing the treatment technology and forging links with the medical device industry.

The group these enthusiasts formed came to be known as the Proton (later changed to 'Particle' as other charged particles such as helium, carbon, neon, and silicon began to be used in some treatments) Therapy Cooperative Group (PTCOG). It was certainly an interesting choice of name, evoking as it did 'cooperative group' moniker long associated the clinical trial groups established under the aegis of the NIH and NCI. From its inception PTCOG was designed to be different to the older, 'official' NCI-sponsored cooperative group on radiation medicine, RTOG, in that they were focused primarily on a single modality: particle-beams. Members of the new group were first and foremost focused on early adoption of technology rather than, say, like RTOG on the organization of clinical trials. The first edition of PTCOG's newsletter, *Particles* (edited by the Harvard cyclotron laboratory biophysicist, Janet Sisterson), outlined this commitment to nurture to a growing community of 'those interested in proton, light ion and heavy charged particle radiotherapy'. A whiff of early adoption proselytizing was also on display in the first newssheet, with comments that *Particles* could and would serve as a resource for those wishing to 'inform radiation oncologists, neurosurgeons, ophthalmologists and others' of the benefits of more widespread use of the modality.[39] In its early days the newsletter kept a still relatively small and tight-knit group of few dozen interested oncologists, physicists, and vendors, abreast of the news from national and international meetings and provided updates concerning any new construction of new proton therapy facilities in the US or abroad. Membership would soon begin to expand significantly. *Particles*, no. 10 (July 1992) recorded circulation figures at just under five hundred, up from the one hundred or so subscribers from the first edition five years before. *Particles*, no. 11 (January 1993), noted that around sixty per cent of newsletter recipients were in the US, a sign of the strong domestic interest in the development of the new technology.

A year after the launch of PTCOG a new US-based organization—The National Association of Proton Therapy (NAPT)—was formed with the explicit mission to promote the new technology. NAPT was the brainchild of James Slater, the radiation oncologist who oversaw the building of the first hospital-based proton therapy centre at California's Loma Linda University, and a former spokesman from the Department of Energy, Leonard Artz. In a later interview, Artz described how he and Slater had first met during their tenure at the Department of Energy-funded Fermi

National Accelerator Laboratory (Fermilab) during the late 1980s. While Slater was at work on a project to commercialize proton therapy technology for the private sector, Artz was the senior press officer in charge of high-energy physics and nuclear medicine for the federal laboratory.[40] These were skill sets that would transfer well into the new organization. As described in *Particles*, no. 8 (June 1991), NAPT was setup with the mission of spreading the message of the 'proton advantage' to Congress, professionals, and the public.[41] During his more than two-decade long tenure at the helm of the NAPT, Artz put his media skills to good use by ensuring a steady flow of upbeat stories for press consumption about proton therapy, all the while scanning for and responding to perceived inaccuracies and negative comments in the media.[42]

No amount of persuasive flair could, however, overcome the fact that any collaboration between Fermilab and the Loma Linda University depended on money, and lots of it. This is where the California Congressman Jerry Lewis stepped in by appropriating $80 million for Fermilab to design and build a proton therapy facility at the school's medical centre. It was a cause that the statesman would support for the rest of his life. In the years following that initial massive injection of funds, Lewis was successful in earmarking a lot more money for Loma Linda. Such earmarking for academia was a controversial practice. For its critics, earmarking was a corrupting manoeuvre that improperly sidestepped academic scrutiny and peer-review. For advocates of appropriations, though, earmarks served to ensure that the less well-known universities were not starved of cash that would otherwise simply congregate within a handful of the more prestigious, and less needy, institutions. In any case, the scale of Lewis' lobbying work on behalf of Loma Linda was quite stunning. During the late 1990s and early 2000s, a time when the congressman served on both the House Appropriations Committee and the Defense Appropriations Committee, the small Seventh-day Adventist school received more government money than any other US university by a wide margin.[43] Little wonder then that Loma Linda was on occasion referred to as 'Loma Lewis University' by political wags.[44]

The Californian facility accepted its first proton therapy patient in 1990, and privately funded ventures in Houston, Texas and Jacksonville, Florida soon followed.[45] Two stimuli were instrumental in changing attitudes about the economic viability of constructing these new proton centres. The first was the 1988 approval by the FDA of the Loma Linda and Massachusetts General Hospital (MGH) proton devices; while the second was the successful application by those same institutions to the American Medical Association (AMA) for the creation of proton therapy 'procedure codes' in 1990.[46] FDA approval opened the procedural pathway for vendors to develop new technologies, while AMA codes provided the route to treatment-reimbursement by Medicare and private health insurers. Taken together, then, the lower initial costs and a more obvious route to treatment reimbursements provided enough of an incentive for manufacturers and providers to be tempted into the marketplace.

Despite these major shifts, the high cost of proton therapy as compared to other kinds of therapy—like, for instance, IMRT—was a persistent problem. Critics seized on the expense of proton therapy just as surely as they did on its steady spread into routine practice in the absence of supporting data from prospective clinical trials. To assess these criticisms appropriately, though, we must first ask whether proton therapy *was* particularly exceptional in its development and uptake in an age of highly expensive high technology medicine.

PROTON THERAPY: A *SINGULARLY* EXPENSIVE, 'UNPROVEN' HIGH TECHNOLOGY?

In 1995 Dan Feldstein, a journalist with *The Houston Chronicle*, published a series of articles critical of the MDACC's new Proton Therapy Center, the third such facility to be built in the United States and the first to be a for-profit enterprise.[47] The $125 million cost of the Texas centre had been raised primarily by local investment banking firm Sanders Morris Harris joining with other regional and national venture capitalists. More controversially, the local police and fire pensions scheme had been encouraged by Sanders Morris Harris to invest $37 million in the scheme, money that was ultimately underwritten by the taxpayers of Houston.[48] With few other proton therapy units existing at the time, and no other for-profit centres in existence, Feldstein reported on concerns about the future sustainability of the treatment, and the place of profit-driven medical technology in the healthcare industry:

> Top doctors at M.D. Anderson view the center, the nation's third major proton facility, as a bold vision worthy of one of the nation's top cancer centers. But the story behind it is as complex as it is visionary, and raises significant questions about medicine, money and the free market for health care technology.

For one, the treatment is substantially more expensive than other treatments, and its advantages have been proved on only a small number of cancers. Its value in treating others, given its cost, is openly debated among experts. Secondly, though the new facility will say 'M.D. Anderson,' it will be owned by a group of private investors who stand to make—or lose—the most money. These private investors have committed millions of dollars and include some of the biggest names in Houston, as well as the Houston firefighter and police public pension fund. For its part, M.D. Anderson has promised to 'promote' proton therapy, for which it could earn a seven-figure bonus if the investors make enough profit. According to contracts obtained by the Houston Chronicle under the Texas Public Information Act, M.D. Anderson is contractually obligated to its venture partners to 'advertise and promote' the therapy, doing things such as touting its benefits on the internet. The proton center estimates that one-third of M.D. Anderson's patients will be suitable.[49]

Responding to criticisms surround the 'promotion' of proton therapy, the then President of the MD Anderson Cancer Center, John Mendelsohn, assured Feldstein of the ethical grounds for the institution's support:

'We are not promoting this in the hyped-up sense. We're going to be educating,' said M.D. Anderson President Dr John Mendelsohn. 'I'm convinced that proton therapy is at least as good as standard radiation therapy, and there are preclinical and scientific data to say it's very possible it will be better,' Mendelsohn said. 'And that is what Anderson is here for.' Aggressive treatments for all patients is 'part of our national culture,' Mendelsohn said, even if it sometimes drives up patient costs.[50]

As discussed in Chap. 6, this 'national culture' of aggressive intervention was certainly on display with the rise in PSA screening during the early 1990s, and it's unsurprising that the building of a new for-profit proton centre coincided with the PSA-inspired prostate epidemic. The promises that proton therapy could not only eradicate the tumour, but to also reduce the risk of incontinence and erectile dysfunction (in comparison with other treatments), held obvious appeal to this, usually younger, new pool of patients. In the cancer business the ability to create such assurances in the mind of the consumer offered an enormous edge over competitors. In the absence of clear indications that proton therapy does any *harm* or has performed less effectively than IMRT, it seems understandable that even the most well-informed men, those able to rise above the marketing and hype, might well just decide to err on the side of precision.

The high costs associated with this form of treatment have continued to stir the controversy. Over a decade after Feldstein's initial reporting a 2009 *New York Times* piece by the economics journalist David Leonhardt summarized the ongoing problems. In particular he highlighted the cost comparison between proton therapy, conservative European-style 'watchful waiting' (or 'active surveillance' as it is more commonly known in the US), and IMRT:

> Some doctors swear by one treatment, others by another. But no one really knows which is best. Rigorous research has been scant. Above all, no serious study has found that the high-technology treatments do better at keeping men healthy and alive. Most die of something else before prostate cancer becomes a problem. 'No therapy has been shown superior to another,' an analysis by the RAND Corporation found. Dr. Michael Rawlins, the chairman of a British medical research institute, told me, 'We're not sure how good any of these treatments are.' When I asked Dr. Daniella Perlroth of Stanford University, who has studied the data, what she would recommend to a family member, she paused. Then she said, 'Watchful waiting.'
>
> But if the treatments have roughly similar benefits, they have very different prices. Watchful waiting costs just a few thousand dollars, in follow-up doctor visits and tests. Surgery to remove the prostate gland costs about $23,000. A targeted form of radiation, known as I.M.R.T, runs $50,000. Proton radiation therapy often exceeds $100,000.[51]

Like proton therapy, IMRT had also faced criticism for being introduced to market with a sparse understanding of its advantages and weaknesses as gleaned through prospective RCTs. IMRT had also spread very rapidly in the worldwide radiotherapy market after its debut in the late 1990s.[52] Within five years some seventy per cent of US radiation oncologists were using IMRT, in spite of lack of trial data and its significantly higher costs as compared to existing standards of care.[53] Proton therapy was not then a unique case in the recent history of radiotherapy for cancer patients. As an x-ray based technology, however, while IMRT was more expensive than conventional radiotherapy, it was still nowhere near as expensive as proton therapy. As I discuss below, advocates and opponents of proton therapy argued from a diverse base of premises, but the high initial cost of the technology and the subsequent high cost of treatment remained the most persistently visible and controversial dimension of the debate.

Proton Therapy and the Controversy Over Clinical Equipoise and Clinical Trials

The exchange between the US oncologists, Harvard Medical School's Michael Goitein and MDACC's James Cox, and the UK's Institute of Cancer Research radiation oncologist Michael Brada (supported by European data analyst colleagues) quoted at the beginning of this chapter, was quite representative of such debates as they were settling around the proton therapy issue at the end of the 2000s. Believing that commercial interests were of central importance to the rapid growth of unproven technologies like proton therapy, Brada's team argued that the high financial outlay necessary for the introduction of new, complex medical technology stimulated and nurtured a spirit of high confidence and enthusiasm:

> The necessary prerequisite for introduction of such technologically complex treatment into the clinical arena is enthusiasm for particle therapy, a belief in its benefit, and considerable financial outlay. The investment in clinical facilities offering proton therapy should not simply follow enthusiasm and belief in the new technology but should be firmly based on objective outcome data demonstrating the real additional value of protons over photons using the criteria of evidence-based medicine.[54]

In a vigorous response, Giotein and Cox not only pointed to the similarities between protons and IMRT in lack of initial RCT data (also quoted at the beginning of this chapter), they also struck back at Brada and other critics from the basis of ethical inquiry. Multi-arm clinical trials, they said, required a core uncertainty to be present in the minds of the investigator as to which treatment arm of the trial was best—a concept known as 'clinical equipoise':

> It is … hard to imagine how any objective person could avoid the conclusion that there is, at the very least, a high probability that protons can provide superior therapy to that possible with x-rays in almost all circumstances. It is primarily for this reason that the practitioners of proton beam therapy have found it ethically unacceptable to conduct RCTs comparing protons with x-rays. Such a comparison would not meet a central requirement for performing RCTs, namely that there be clinical equipoise between the arms of the trial.[55]

As a self-styled 'protoneer',[56] we might expect Goiten to be energetic in opposition to anything that could be perceived as roadblocks to the progress of proton therapy. He was certainly joined in this 'call to equipoise' by Cox in Houston, obviously, but also by other senior radiation oncologists like Herman Suit at MGH.[57] Clinical equipoise, as the originator of the term Canadian bioethicist Benjamin Freedman defined it, rested on the notion of uncertainty. Without uncertainty, Freedman argued, there could be no *ethical* clinical investigation:

> [I]t is necessary that the clinical investigator be in a state of genuine uncertainty regarding the comparative merits of treatments A and B for population P. If a physician knows that these treatments are not equivalent, ethics requires that the superior treatment be recommended. ... Equipoise is an ethically necessary condition in all cases of clinical research. In trials with several arms, equipoise must exist between all arms of the trial; otherwise the trial design should be modified to exclude the inferior treatment.[58]

In their response to Brada then it was this inability *to be uncertain* that took on central importance for Goitein and Cox. Other critics of Goitein and Cox took a more moderate view than Brada and recognized the equipoise concern but argued that it might only apply in limited cases. Here the argument went that while clinicians might reasonably lack equipoise in consideration of trials for, say, paediatric tumours where late adverse reactions were of huge concern, such a case was much more difficult to make for, say, the treatment of older men with prostate cancer.[59] In 2009 an Agency for Health Research Quality (AHRQ)-funded systematic review, a team from the Evidence-based Practice Center at Tuft's Medical Center, Massachusetts, led by Thomas Trikalinos, considered the issue of equipoise. They argued that while there were likely *some* cases of concern for clinical equipoise in proton therapy research in some rare cancers, there were very significant doubts as to superior therapy in more common cancers, such as those of the prostate. In other words, clinical equipoise in that latter case could indeed be achieved.

The AHRQ's report on particle beam radiation was concerned primarily with questions other than equipoise though, commissioned as it was as part of the Agency's 'Effective Healthcare Program'—an EBM inspired effort launched in 2005 to provide comparative effectiveness assessments of different treatment options.[60] In preparing the report, the Tufts team analysed the literature on all types of particle therapy for cancer. They found that, while a number of charged particle treatments were in use, the vast majority of patients around the world treated with charged particles received them in the form of proton beams (eighty-seven per cent—the remaining share being mostly made up by helium or carbon ion treatments).[61] The group also reported that the literature on particle therapy, while large, was little concerned with RCTs comparing fast particles to other types of radiotherapy, finding that only 'a handful of RCTs and non-randomized comparative studies were identified, and they compared lower vs. higher doses of particle beam therapy, particle beam therapy alone vs. other treatment, or incorporation of particle beam therapy to a treatment strategy vs. not'.[62] They were mostly studies that did not, in other words, seek to investigate whether particles were *comparatively* more effective than other modalities.

While the general lack of comparative research was troubling to the team, they were particularly critical of the lack of RCTs given, as they said, the 'numerous examples of interventions that, despite very favorable and strong pathophysiological rationale, turned out to be harmful when evaluated in RCTs'.[63] A nod back here, then, to the protoneers and their appeal to the pathophysiological rationale as a justification to press on with the expansion of proton therapy as treatment modality in prostate cancer. Another major review of proton therapy, this time using Surveillance, Epidemiology, and End Results (SEER) data from the NCI, appeared in a 2012 special issue of the *Journal of the American Medical Association* devoted to comparative effectiveness research.[64] Once again an exhaustive review came up empty on definitive proof of the benefit of protons over IMRT in the treatment of prostate cancer.

It was not, it should be said, necessarily the case that proton therapists as a group *were* particularly unwilling to engage in comparative trials. Opportunities for such engagement were, after all, rather limited and this fact might just as well explain the scarcity of trials. James Cox (he of the equipoise argument described above), discussed the chronic lack of funding for trials of new technologies in a 2008 editorial regarding the use of 'CyberKnife'—a robotic radiotherapy delivery system developed by the US medical device manufacturer, Accuray.[65] 'The U.S. National Cancer Institute is disinclined to fund comparative trials of technological advances',[66] wrote Cox. Acknowledging that vendors might take up a role in funding comparative trials—in much the same manner as pharmaceutical companies did in drug trials— he pointed out a major wrinkle of this approach: 'Accuray has proposed such support of CyberKnife studies but has rejected the suggestion that other vendors with similar technologies participate.'[67] While specific in this instance to Accuray and Cyberknife,

Cox's comments were suggestive at least of how such proprietary concerns might have hindered other vendor-sponsored trials for technology assessment.

Another aspect of this debate worth mentioning is that assessments for medical devices were, by design, very different to those for drugs: RCTs, it should be remembered, were developed and used primarily to test drugs, not devices. The radiation oncologist Søren Bentzen pointed out some of the ways technology assessment in the US differed from trials of drugs, and commented on how such differences were regarded by the medical community at large:

> An interesting asymmetry exists between getting approval for a drug compared with a medical device. The FDA approves a new drug for a given medical indication based on evidence from randomized controlled trials that the drug provides a net benefit over standard therapies. By contrast, medical devices can be marketed with a so-called 510(k) FDA approval; in essence a certification that states that the device does what it is meant to do and that using it does not compromise patient safety. Critics argue that a new technology is 'just another drug' and, therefore, the benefit of any new device should be demonstrated in randomized controlled trials before FDA approval.[68]

While sympathetic in principle to the pursuit of evidence by means of clinical trials, Bentzen warned against insisting on such 'purity' in practice. Quoting Voltaire's maxim that 'the best is the enemy of the good', he made his appeal for pragmatism:

> Maybe in the case of health technology assessment, Voltaire was right; if we insist on the 'best', namely randomized comparisons of treatment outcome from new technologies—a bar raised so high that in practice we rarely reach it—we will continue to miss out on the 'good', namely critical, systematic comparisons of technologies and devices in terms of operational or quality criteria.[69]

Bentzen concluded his editorial with a further appeal to the realities of trying to apply new devices in the absence of data from rigorous trials (and a switch in literary illusion): 'Radiation oncologists must engage in the development of novel paradigms for critical health technology assessments without the ideal requirement of randomization. Let us start doing what we can, rather than continuing to wait for Godot!'[70]

This attitude of cautious optimism in the utilization of new technologies might well have annoyed clinical trial advocates, but the successes of IMRT had buoyed the confidence of oncologists like Bentzen. The fact remained, however, that RCTs could and often did throw up problems that no one had previously noticed or anticipated. While not itself a report of a clinical trial, the 2012 SEER-data report by Nathan Sheets and others, for instance, uncovered a pattern of reporting on some unexpected problems: specifically, a higher incidence of gastrointestinal complications in men treated with proton-beams as compared to IMRT (something they speculated might be to do with the difficulty of keeping patients absolutely still during treatments).[71]

Following the AHRQ report, the NCI did finally act. In 2012 the MGH along with seven other providers of proton therapy launched an NCI-sponsored phase III multi-site RCT to compare proton therapy with IMRT—the Prostate Advanced Radiation Technologies Investigating Quality of Life (PARTIQoL) study.[72] At the time of writing, patients are still being enrolled into the PARTIQoL, and early results are unlikely to be reported until 2018. PARTIQoL is interesting for other reasons too. It brings together famous teaching hospitals, like MGH and MDACC, long involved in particle research and therapy, with more recent 'entrants' to the field. This latter group includes the Central DuPage Hospital, part of a private provider network owned by the Northwestern Medicine Regional Medical Group, and ProCure, the operator of a network of proton therapy centres founded by the nuclear physicist John Cameron (a figure heavily involved in the early development of the modality while working with the cyclotron at Indiana University). The makeup of the PARTIQoL group is highly representative of the radiotherapy marketplace as it exists in the early twenty-first century, and it is worth considering how this provider pluralism first emerged.

THE INFLUENCE OF PROSTATE CANCER ON THE GROWTH OF PROTON THERAPY

While evidence supporting the use of proton therapy in prostate cancer was tenuous at best in the 2000s, this was not the case for other applications of the modality. A number of clinical studies published at that time reported positive results in the use of proton beams for ocular tumours and some paediatric cancers.[73] The fragile structures of the eye, and the close proximity of tumours to (highly radiosensitive) developing organs in the paediatric patient, sometimes made surgical intervention either very difficult or simply impossible; in these cases it appeared that proton beams might in fact become the superior treatment of choice. Whatever the case, though, these categories of cancer were relatively rare in their occurrence, and such limited patient populations would not fill the capacity of existing facilities, let alone the additional centres planned for the 2010s.[74] Prostate cancer was obviously a very different matter: higher rates of screening had led to higher rates of detection in the 1990s and beyond, and this had, in turn, boosted the size of potential treatment populations in what was an already comparatively common cancer. Due to the sheer size of the patient group, attracting prostate cancer patients would become a vital part of the growth of proton therapy centres in the United States.

Loma Linda is a good example of a facility sustained through the treatment of prostate cancer patients. Within three years of opening in 1990, the Californian proton therapy centre had treated six hundred and eighty-two patients, four hundred (or fifty-nine per cent) of who were treated for prostate cancer.[75] Fourteen years later, a progress report by Loma Linda's Chairman of Radiation Oncology, Jerry Slater, noted that while the facility treated some fifty different anatomical sites, carcinoma of the prostate was the most common condition treated. In fact, by the mid-2000s the percentage of patients treated for prostate cancer had risen to sixty-five per cent of all cases seen at the centre.[76] Loma Linda was not alone in this focus. In a 2009 review of the field, none other than Michael Brada used this fact as further evidence for what he regarded as the overly hasty acceptance of proton therapy as a major treatment modality.[77]

Not all proton centres followed this pattern though: the facility at the Mayo Clinic in Minnesota, for instance, kept *its* focus on childhood cancer, for which evidence of efficacy was much more advanced. It's also worthwhile to note, I think, that not all criticism of proton therapy came from EMB enthusiasts, nor was it all from outside of the practice community. A 2012 interview with Robert Foote, a radiation oncologist and proton therapist at Mayo, makes this point clear. An optimist for the future of proton therapy, Foote nonetheless lamented the overly rapid expansion of the modality. It was an expansion, he believed, that was generated and sustained through the questionable channelling of large numbers of patients with prostate cancer into proton therapy centres. Centres that were, in turn, financially reliant on this patient group. Foote did not mince his words when he suggested that a bifurcation in US proton care had essentially created two models of practice, 'one to make money—the other to provide the best care possible for the people who need it'.[78]

This dependency on prostate cancer patients was, perhaps, particularly pronounced in the case of the for-profit proton centres, such as those run by the Northwestern Medicine Regional Medical Group and ProCure. For-profit provider networks like these had first emerged in the US in the late 1980s when physician groups, hospital groups, pharmaceutical companies, health insurance companies, and others bought up hospitals and private practices to form physician management organizations. The purpose of these groups was to centralize administrative and technical services and then either 'lease' these resources back to physicians on a fee-for-service model or else hire physicians (and later nurses) directly as salaried employees or as part of an 'employee bank' covering staffing gaps in area hospitals.[79] While many of these companies failed in the 1990s (mostly due to the lack of physician recruitment), some thrived: typically those which focused on a particular kind of specialist practice where access to the resources of a large network made the most sense to individual practitioners.

When organizational growth of for-profit networks did occur, moreover, it was often at the expense of the more established academic medical centres. Take, for instance, one of the original surviving private provider networks, Texas Oncology, which as part of its umbrella group US Oncology, became by the late 1990s the single largest provider of cancer care in the nation.[80] It was an important feature of Texas Oncology's success that they were able to divert profitable patients away from the MDACC, in large part due to their ability to provide the services patients wanted within their own communities.[81] While MDACC did try to compete with US Oncology in Texas by forming its own comprehensive community-based cancer program in the mid-to-late 1990s, it was to decidedly mixed results. Like Texas Oncology, ProCure was based on the gamble that fee-generating patients could be attracted in large enough numbers to turn a profit. Like Texas Oncology, ProCure also stressed its ability to bring high technology therapy much closer to the patient's own home.[82] And finally, like Texas Oncology, ProCure ensured that it owned the most profitable parts of the whole cancer treatment business, namely, radiation oncology treatment centres.

The question remains, though, of *why* so many men with prostate cancer sought out proton therapy treatment in the 1990s and 2000s. Part of the answer is likely to do with referral patterns: notwithstanding questions surrounding financial interest on the part of physicians, the improved geo-graphic access to proton therapy would seem to be an important factor in treatment decision-making for patients. Indeed, a group of urologists and public health researchers associated with the University of California system and the Kaiser Permanente Medical Group (a not-for-profit man-aged care organization well-known for its health services research) made a convincing case for this in their 2012 article, Proton Beam Therapy and Treatment for Localized Prostate Cancer: If You Build It, They Will Come.[83] Provider networks have for years tried to ensure that they stop patients being 'referred out' of the network and on to a different provider by acquiring radiotherapy facilities of their own and placing them close to their patients. Texas Oncology opened a radiotherapy centre in the small town of Amarillo Texas in 2007, for instance, and, in the early 2010s it became involved in the construction of the Texas Center for Proton Therapy located in Irving, a suburb of Dallas (bringing the number of proton centres in this single state to two). The spread of proton therapy to the community setting has depended on technological advances like the scaling-down of cyclotrons, but is also a phenomenon driven by business practices like those of Texas Oncology and ProCure that have sought to design 'turnkey' operations, freed from the need of the support of a large multi-disciplinary academic medical centre.[84]

Whether as a result of proximity or not, it seems very clear that many of the men who were treated with proton therapy actively sought the treat-ment of their own volition. In their SEER study, Sheets and colleagues linked the growth in proton facilities to vigorous direct-to-consumer advertising, especially around the issue of precision treatment.[85] As the public health researcher Julie Donohue points out in her history of direct-to-consumer advertising, the 1990s was a potent time of increased patient awareness.[86] Trends in consumer rights and information technology had by then converged to produce patients motivated and able to seek out health information and support from a variety of sources, not just those provided to them in the doctor's office. What really changed in the late 1990s, however, was the astonishing injection of advertising money into the medical marketplace after the FDA relaxed restrictions on prescription drug and other kinds of medical advertising in 1997.[87] Medical advertis-ing in the US is as old as the country itself, of course. As we saw in the discussion of testicular extracts in Chap. 3, by the interwar years of the twentieth century the peddling of proprietary wares by 'entrepreneur-ial' doctors faced the increasing ire of a medical establishment who dismissed them as 'quacks' and accused them of bringing the profession into disrepute. When the approximately forty-seven million dollars spent on medical advertising in 1990 had by the year 2000 blown up into two-and-a-half *billion* dollars,[88] old objections were stirred. Critics argued that the investment of such enormous sums of money had the potential to threaten the health and safety of the consumer-patient if financial interests were allowed to influence the medical message by exaggerating benefits while minimizing possible risks.[89]

In 2006 *The Oncologist* published the transcript of a MGH-Schwartz Center Rounds (a monthly multidisciplinary forum for caregivers and patients) that took up the issue of direct-to-consumer advertising. Was, the panel wondered, direct-to-consumer advertising in oncology different from other kinds of medical promotion? Well, yes, argued an oncology fellow, 'oncology patients and their families are especially vulnerable to direct-to-consumer advertising because they are desperate to learn about any therapy that might possibly work ... [T]here is still no other diagnosis that is met with such trepidation and desperation.'[90] Well, maybe, replied a patient in remission from ovarian cancer. Direct-to-consumer advertising had the potential to help patients to be more informed she said, although not always. Advertisements pitched to cancer patients, she had observed, tended to be ones in which 'the bad news is airbrushed away [and i]t is left to the physician to deliver the bad news that the ads carefully omit'.[91] So what was the attraction for patients in direct-to-consumer advertising? 'Mostly, it's an effort, I think to right the balance of power in a situation in which we feel powerless.'[92] My opinion, said a senior oncologist, 'is that the mission of direct-to-consumer advertising is to make a profit, artfully disguised as providing a service and educating'.[93]

LOOKING TO THE FUTURE: THE CLINICAL INVESTIGATION AND PRACTICE OF PROTON THERAPY

The proton therapy story shows us how novel medical interventions entered and changed the market place in the late twentieth century. Nurtured in the incubator of the 'Big Science' of post-WWII medical physics research, particle therapy had jumped the walls of academia by the turn of the century—thanks, in large part, to men with prostate cancer. In the absence of supporting data from clinical trials, commercial promotion, professional enthusiasm, and consumer demand, were all potent indicators of marketplace success. In recent years, the commercial practices of Texas Oncology, ProCure, and organizations like them, have attracted disapproval. Critics charge that the 'skimming' of lucrative patients by for-profit networks has often left not-for-profit facilities with a proportionally higher load of loss-making, low-income patients.[94] As institutions in receipt of taxpayer dollars, the effects of this redistribution has the potential to reverberate across the healthcare system and beyond.

Not all trends point in favour of the proton therapy industry, however. In 2013 the US Government Accountability Office (GAO) used a review of Medicare data to issue a report to Congress on the rumoured trend of clinicians sending patients to treatment facilities in which they themselves had financial interests. It was an issue that was then in the public eye, thanks to some high profile media attention.[95] The GAO report did find some major problems. It appeared that 'self-referring' providers (meaning those who had some financial interest in the treatment facility to which their patient was sent) were much more likely to prescribe more costly treatments than other comparable providers with no such financial interests.[96] While the report did not mention proton therapy at all—it was, in fact, focused on the data concerning IMRT referral patterns—it seems unlikely that proton therapy will escape similar suspicion and scrutiny in the future.

The proton therapy story might not necessarily continue to be one of growth for other reasons too. In December of 2014, the Indiana Proton Therapy Center permanently closed its doors, with an independent review of the closure citing a range of contributing factors, including the prohibitive costs of maintaining an aging cyclotron. Beyond the problems associated with the specific centre, however, the reviewers also raised the considerably more troubling possibility that the wider industry might be characterized as a living in a 'proton bubble'.[97] With a new reluctance on the part of major private health insurers—Aetna and Cigna among them—to continue to cover proton therapy for prostate cancer patients, it is easy to see how the bubble might be on the verge of bursting. Without the income from prostate cancer treatment, it is difficult to see how proton therapy facilities can remain viable. It remains to be seen, of course, what impact the findings of PARTIQoL will have on these questions and concerns. How will proton therapy providers, and indeed future patients, respond to a new age of evidence-based practice in proton care?

As I have argued in this and previous chapters, this story of prostate cancer treatment is also a story about the nature of evidence in medicine and the sometimes tenuous relationship between that evidence and actual clinical practice. Scholarly histories of post-WWII biomedicine in the US, perhaps particularly in the field of oncology, tend to be generally enthusiastic about the rise of the randomized clinical trial as the gold standard of clinical research. By contrast, there is relatively little historical attention focused on the decades of critical commentary concerning clinical trials, criticisms that litter the editorial, review, and letters pages of most major medical journals. Historical coverage of the successes and failures of the clinical trial is at best asymmetric and this is important. A focus on success implies that over time medical practice has become increasingly evidence-based. As Iain Chalmers and others have argued, however, the systematic implementation of fair trials in medicine has been, and remains, a depressingly elusive goal.[98] At a broad stroke, we might worry with good reason that industry-funded research is more likely to show bias towards the publication of positive results,[99] while the government bureaucracy surrounding federal grants leads researchers into years of frustrating red-tape, with trial data frequently abandoned or if the data are published, published against a moving background of ongoing bioscientific research.[100] The gold standard might not be as securely anchored in the future of twenty-first century medicine as historians and social scientists seem to assume.

NOTES

1. Brada, Pijls-Johannesma, and Ruysscher, Proton Therapy in Clinical Practice, 968.
2. Goitein and Cox, Should Randomized Clinical Trials Be Required for Proton Radiotherapy?, 176.
3. Porter, *The Greatest Benefit to Mankind*, 607–8.
4. Young, *Technique of Radium Treatment of Cancer of the Prostate Gland*, 3.
5. Ibid., 34–6.
6. Aronowitz, Whitmore, Henschke, and Hilaris, 157.
7. Ibid.
8. Machtens et al., The History of Endocrine Therapy of Benign and Malignant Diseases of the Prostate, 224–5.
9. Rosevear et al., Rubin H. Flocks and Colloidal Gold Treatments for Prostate Cancer.
10. London Health Science Center, Celebrating the 60th Anniversary of the World's First Cancer Treatment with Cobalt-60 Radiation.
11. Ueyama and Lecuyer, Building Science-Based Medicine at Stanford.

12. See Bagshaw, Kaplan, and Sagerman, Linear Accelerator Supervoltage Radiotherapy, 127.
13. Ibid., 128.
14. Mazzitelli, Application of Particle Accelerators in Research, 372.
15. Yates, Vaessen, and Roupret, From Leonardo to Da Vinci, 1710.
16. Heilbron and Seidel, *Lawrence and His Laboratory*, 229.
17. Ibid., 219.
18. Stone and Larkin, The Treatment of Cancer with Fast Neutrons.
19. Ibid., 613.
20. Ibid., 612.
21. Ibid., 614.
22. Halperin, Particle Therapy and Treatment of Cancer, 677.
23. Fowler, 40 Years of Radiobiology, 103.
24. Duncan et al., Fast Neutrons in the Treatment of Head and Neck Cancers.
25. Jones, The Neutron-Therapy Saga.
26. Kacperek, Protontherapy of Eye Tumours in the UK, 378.
27. Cox, Evolution and Accomplishments of the Radiation Therapy Oncology Group, 747.
28. Ibid., 748.
29. Ibid.
30. Griffin, Fast Neutron Radiation Therapy, 18.
31. Olson, *Making Cancer History*, 200.
32. Griffin, Fast Neutron Radiation Therapy, 19.
33. Hall, Protons, 9.
34. Ibid., 196.
35. Smith, Vision 20/20.
36. Mackie, History of Tomotherapy, R427.
37. Metz, History of Proton Therapy.
38. Bucci, Bevan, and Roach, Advances in Radiation Therapy, 117–18.
39. Particles.
40. Ruck, Q&A with Leonard Arzt.
41. PTCOG, Particles.
42. The National Association for Proton Therapy, Community Forum.
43. Kammer, Lobbying Ties Give Campus Funding Edge.
44. Kammer, A Steady Flow of Financial Influence.
45. Smith, Vision 20/20, 560.
46. Ibid., 559.
47. Feldstein, Proton-Therapy Costs vs. Benefits.
48. Feldstein, Police, Fire Pensions Warned about Deal.
49. Feldstein, Proton-Therapy Costs vs. Benefits.
50. Ibid.
51. Leonhardt, Health Reform's Acid Test.

52. Terasawa et al., Systematic Review, 559.
53. Veldeman et al., Evidence behind Use of Intensity-Modulated Radiotherapy.
54. Brada, Pijls-Johannesma, and Ruysscher, Proton Therapy in Clinical Practice, 965.
55. Goitein and Cox, Should Randomized Clinical Trials Be Required for Proton Radiotherapy?, 175.
56. Goitein and Goitein, Swedish Protons, 793.
57. Suit et al., Should Positive Phase III Clinical Trial Data Be Required before Proton Beam Therapy Is More Widely Adopted?
58. Freedman, Equipoise and the Ethics of Clinical Research, 141.
59. Lewis, On Equipoise and Emerging Technologies.
60. Trikalinos et al., Particle Beam Radiation Therapies for Cancer.
61. Ibid., 9.
62. Ibid., 33.
63. Ibid., 34.
64. Sheets et al., Intensity-Modulated Radiation Therapy.
65. Cox, Assessment of Technological Advances in Radiation Oncology.
66. Ibid., 11.
67. Ibid.
68. Bentzen, Radiation Oncology Health Technology Assessment, 563.
69. Ibid.
70. Ibid.
71. Sheets et al., Intensity-Modulated Radiation Therapy, 1619.
72. Efstathiou, Proton Therapy vs. IMRT.
73. Olsen et al., Proton Therapy.
74. Kagan and Schulz, Proton-Beam Therapy for Prostate Cancer, 408–9.
75. PTCOG, Particles.
76. Slater et al., The Proton Treatment Center at Loma Linda, 85.
77. Brada, Pijls-Johannesma, and De Ruysscher, Current Clinical Evidence for Proton Therapy, 321.
78. Epstein, Is Spending on Proton Beam Therapy for Cancer Going Too Far, Too Fast?, e2488.
79. Burns and Robinson, Physician Practice Management Companies, 4.
80. Fintor, For-Profit Treatment Centers, 1272.
81. Lau, U.S. Oncology Inc.
82. Procure, Learn About ProCure.
83. Aaronson et al., Proton Beam Therapy.
84. Harvey, Proton Therapy.
85. Sheets et al., Intensity-Modulated Radiation Therapy, 1618.
86. Donohue, A History of Drug Advertising, 682.
87. Ibid., 685.
88. Fintor, Direct-to-Consumer Marketing, 330.

89. Ibid.
90. Abel et al., Direct-to-Consumer Advertising in Oncology, 218.
91. Ibid., 219.
92. Ibid., 220.
93. Ibid., 221.
94. Hessel, Cancer Capitalists.
95. Gawande, The Cost Conundrum.
96. Government Accountability Office, Medicare: Higher Use of Costly Prostate Cancer Treatment.
97. Lee, As a Proton Therapy Center Closes, Some See It as a Sign.
98. Evans, *Testing Treatments*.
99. See, for example, Lexchin, Those Who Have the Gold Make the Evidence.
100. Nass, Moses, and Mendelson, *A National Cancer Clinical Trials System for the 21st Century*, ix.

CHAPTER 8

Conclusions: Medicine, Masculinity, and the Problems of the Prostate

Currently, it is becoming clear that the prostate gland may be the major site for medical problems in the American man. Abnormalities in prostate growth and infection in human prostate glands produce some of the most common, costly, and devastating disease occurring in men.

Donald Coffey, Prostate Cancer: An Overview of an Increasing Dilemma (1993)[1]

The twentieth century decline in the incidence and prevalence of infectious disease has long been recognized by historians as coinciding with a renewed biomedical focus on the 'disease management' of the chronically ill. During the 1960s and 1970s the use of 'risk factors'—clinical indicators, genetic markers, lifestyle choices, and the like—began to increase the frequency and intensity of similar disease management interventions in seemingly healthy populations. During the past forty years the global healthcare industry has engineered hugely profitable markets from healthy 'patients', largely by appealing to the value of preventative intervention in the battle against the new diseases of civilization: hypertension, cancer, and diabetes. Robert Aronowitz,[2] Ilana Löwy,[3] and Charles Rosenberg[4] have all documented disturbing trends in disease management directed at the aggressive prevention of *anticipated* undesirable outcomes. New diagnostic tools and larger programs of more biologically sensitive screening have led to ever greater 'early detection' of 'pre-cancerous', 'pre-diabetic', and 'pre-hypertensive' patient populations. As Aronowitz points out,

© The Editor(s) (if applicable) and The Author(s) 2016
H. Valier, *A History of Prostate Cancer*,
DOI 10.1057/978-1-137-56595-2_8

the experiences and patient-pathways of these 'pre-patient' patients can become almost indistinguishable from those patients with serious clinical symptoms of disease.[5] While the consequences of this elision between statistical *risk* of disease and actual organic illness can be relatively benign, Aronowitz, Löwy, and Rosenberg highlight at least one dire consequence of this trend: the rising number of healthy but 'BRCA positive' women undergoing extremely drastic measures such as prophylactic double mastectomies. In this book I have made similar observations about similarly drastic interventions in men showing prostatic malignancy as a result of biopsies driven by the PSA explosion of the early 1990s.

The idea that developers of medical technologies and pharmaceuticals 'look for' likely patient populations to diagnose and treat is far from new.[6] Of greater interest perhaps are the numerous case studies that, when taken on aggregate, seem to suggest that new disease categories might be *routinely* created from the stuff of abnormal cervical smears, mammograms, and blood tests. That the application of new medical technology *routinely* shifts disease management policies designed for the treatment of advanced disease to earlier and earlier 'stages' of (pre)disease states is a startling and perplexing claim. In this book I have added to the arguments of earlier authors who proposed that disease screening can 'create' a pre-disease state (in this case, asymptomatic prostate cancer) and thus invoke a disease management intervention. Similarly, I have shown that disease 'advocates', be they members of the healthcare industry or patients themselves, organize to increase awareness and resources so reifying the new 'disease' category. The prostate story does add another dimension to this analysis, however, in that an upsurge in new *screening* technology coevolved with a new type of anti-cancer *treatments*—technologies such as robotic surgery and proton beam therapy. While other studies highlight how pre-patients are managed through interventions designed for the seriously ill, in the cases I analysed in this book we see how screening for the pre-patient helped create a *novel* patient pathway, one that would have consequences for the treatment and management of the seriously ill as well as the apparently healthy. Here, we have an example of the pre-patient sustaining the growth of a new treatment modality; a technology that continues to expand even as the plausibility of the notion of the pre-patient *as* the patient has come into dispute. As I have shown, such pathways can emerge with a promised market (in this case, a very commonly diagnosed cancer) and persist due to market and patient enthusiasm.

MEN, MARGINALIZATION, MASCULINITY, AND METAPHOR IN THE LATE TWENTIETH CENTURY

As I discussed in Chap. 1, part of the reason I wrote this book was to provide something that is, in my opinion, long overdue—a book-length account of a common male cancer. By stressing the disparity in scholarly attention to male and female cancers, it is not my intention to replicate anything of the bitter and accusatory tone that characterizes some of the more extreme rhetoric coming out of some of the prostate cancer advocacy groups.[7] The sense of outraged inequity that frames such language is more often than not targeted at breast cancer, for the large amount of research funding it receives and for the hard-to-miss visibility of industry-sponsored (and industry-serving) 'think pink' campaigns, clothing, walks, events, and so much more.[8] The pink ribbon is difficult to avoid.

It is not to my knowledge part of the charge of unfairness and inequality within prostate cancer activism that cancers other than breast cancer contribute to marginalization. Other very high profile cancer patients like children with leukaemia and lymphoma, for example, might be (but aren't) similarly accused of stealing the limelight and the resources away from men with prostate cancer. The intensity of attention paid to breast cancer (to the exclusion of other well-funded cancers) would then seem to rest on a wider assumption about rampant misandry affecting the activities and priorities of politicians, medical researchers and, it would appear, society at large. This is a strange claim for several reasons but an obvious objection to it would be that, given all the ways in which medicine and society have tended to underserve the needs of older people, ageism would be at least as likely as misandry to be responsible for any sidelining of prostate cancer, a disease that is after all overwhelmingly a disease of older men. Although one might never know it from the anti-breast cancer campaigns, which focus almost exclusively on younger women, so too is breast cancer; another indication that there are issues beyond gender of importance here.[9]

Until very recently, historically speaking, medical research was overwhelmingly focused on men because men (especially younger white men) were for centuries assumed to be the biological 'norm' of human life. One well-known example of the kinds of harms that could follow from such an assumption is in public health messaging around cardiac disease. The white-collar middle-aged man who became the poster child of public health messaging of the American Heart Association as it grew to prominence during the 1950s and 1960s, belied the fact cardiac disease was a critical public health issue not only for the white-collar worker but also middle aged blue-collar workers, middle aged women, and the elderly of any social class. The most well-known signs of heart attack—crushing chest pain, pain radiating to the arms, neck, and jaw—are in fact the common signs in men, not women. The prevalence of these signs as indicative of the 'typical' can mean that women in imminent danger of a heart attack will dismiss symptoms of chest pain when other indications (like shortness of breath or nausea) do not fit the classic (male) model. A formal recognition that a biomedical model of health and disease so comprehensively tied to the physiology of white, adult male adults could be skewing data and underserving women and minorities came with the passage of the NIH Revitalization Act of 1993.

In spite of the NIH-directed reform efforts, overall enrolment of minority populations in clinical trials in the years since has been slow.[10] A decade on from the Revitalization Act, there is a reason (and not a good one) why the cardiologist and author Nieca Goldberg chose to write a book on female cardiac health titled *Women Are Not Small Men*.[11] The *status quo* has proved stubbornly resistant to change. More than a decade on from Goldberg's book, and notwithstanding the numerous public health efforts and campaigns that preceded and followed, the messaging around heart attack in women seems to have still not penetrated much into the public consciousness.[12] Such things matter and matter very much, especially since heart disease is the biggest killer of American women (and men). The overwhelming visibility of the breast *might* (that's a topic for another book) be distracting women from diseases of the chest (in addition to heart disease, significantly more American women will die of lung cancer than breast cancer). If such was indeed the case it would certainly turn accusations of misandry on its head.

While the biomedical project of the twentieth century had men squarely at its center, this is true only in so far as it pertains to men as objects of clinical research and therapeutic intervention. The rather less tangible phenomena of subjective masculinity and gender identity are, by contrast, not well integrated into the biomedical model. The notion that a fretful masculinity causes men to be universally reluctant to ask for help (lest they cede control or display weakness) is by this point so often repeated as to be a cliché. It seems unlikely that such sweeping generalizations do much accurately to depict the complex attitudes that men have about themselves and their health. What seems clearer is that creative strategies for patient engagement that incorporate gender identity (especially by moving away from a one-dimensional view of masculinity that reduces it to stubbornness) might improve clinical encounters and clinical outcomes. Prostate cancer activists developed their own versions of gendered advocacy in the 1980s and 1990s. By harnessing the power of that old friend of the public health campaign, the arresting metaphor, images of stoic masculinity nevertheless were turned to an emphasis on self-care. The uneasy transition between these two ideas might well be at the root of why the campaigns chose *such* hyper-masculine metaphors—of sport, of war, of competition of all kinds.

Strident and militaristic metaphors for cancer do not just appear in relation to *men* with cancer, of course. Whether in popular culture, the medical and health promotion literature, or as it is here in the United States, where direct-to-consumer advertising of hospitals and treatment centres abound, on television and radio too, the language of 'warriors' and 'survivors' of an all-out war against cancer is everywhere. These are metaphors embraced by many patients and I would not care to pass judgment on that—consciously embracing fearlessness within a situation that causes fear seems like a very human thing to do after all. However, as numerous scientists, clinicians, and patients have warned, war metaphors can harm as well as help.[13] Metaphors that exist to persuade men to 'act responsibly' or 'take control' of their health by seeking medical advice can be easily repurposed for the message that real men 'fight' their diseases (perhaps even to the 'death') which opens the way to the most aggressive interventions and the belief that that way is the *only* way.

Military metaphors in prostate cancer campaigns are, I think, a little different from the rest to the extent that they are used with such frequency and in such extended detail. The (in)famous 'dog tag' campaign launched by the Canadian Prostate Cancer Network in 2000 is an illuminating example of a metaphor pushed to extremes. The campaign was premised on the distribution of military-style identification tags (emblazoned with the organization's logo) which men would then wear to promote awareness of PSA screening. For the organizers, the tags were intended to symbolize the courage and camaraderie of men under arms. In wearing them, ordinary, civilian men might achieve something of a connection to the iconic masculinity of the solider, a figure simultaneously strong and independent but with the capacity for great brotherly love and interconnectedness. There was, in other words, nothing 'sissy' about solidarity. The campaign came in for almost immediate derision, particularly since, as its critics pointed out, the reason why serving soldiers wore dog-tags was as an aide to identification should they be incapacitated or killed (this latter case being why the tags are worn as a double set, so that one may be left behind with the body to be recovered later if evacuation is not immediately possible). Not the most uplifting or appropriate metaphor then if considered in any detail.

Not all activism in men's health focuses on military metaphors, and if we take the more recent example of the massively popular campaign started by the Movember Foundation we see some new strategies at work (although that 'new' seemed to contain plenty of the 'old'). Originating in Australia in 2004, Movember—a portmanteau of the Australian diminutive for moustache ('mo') with the month of the year in which it campaigns most visibly (November)—began with a mission to raise awareness about prostate cancer (this was later expanded to other male health issues including testicular cancer and mental health).[14] The campaign is based on encouraging men to seek sponsorship to grow moustaches and beards (the more outlandish the better) in the month of November and then to share their personal journey to the hirsute on their social media accounts. Images and stories about the 'Mo Bros' (and the vocal support they receive from their 'Mo Sistas') have saturated social media during every November for the past decade (and they have certainly made their presence known in other forms of media also). It is a remarkable feat of marketing that has made the organization the most visible and most successful prostate cancer advocacy group in history. The Foundation claims to be the largest non-governmental backer of prostate cancer research in the world, and there is little reason to doubt that this is the case given the global reach and popularity of the charity.[15]

As popular and effective at charity fundraising as it might be, and as fun and friendly an approach to cancer awareness as it might appear, Movember's approach is not without problems. The campaigns rest on the hope that the masculinity symbolized by luxurious facial hair might be morphed (just as it was hoped that the masculinity of the solider might be morphed) into an interpretation of masculine 'solidarity' that urges connection and care of the self: physical, spiritual, and emotional.[16] Such aims are laudable, as is the sense of solidarity that might flow from the campaign for men with prostate cancer or those worried about their health and hesitant to seek medical attention. More specifically, the visibility of a proud moustache or beard might, by tapping into this alternate narrative of masculinity, serve to help men brace themselves for a digital rectal exam and to overcome any lingering sense of shame or weakness that they might otherwise feel is also to be applauded. Whichever way a metaphor is packaged, though, it seems that all roads lead to Rome. The savvy marketing of the Movember Foundation might seem a world away from the heavy-handed military messaging of earlier advocacy groups, but the end goal remains unchanged: to encourage men to get screened, early and often.

Movember's screening-as-prevention focus might be usefully widened to include an awareness of what living with the disease (or the consequences of its treatment) might look like. Therein lies a problem, however. The very thing that gives Movember its distinctiveness and its popularity—the growing of beards and moustaches—is also a potential source of alienation for the very men it strives to support. Men undergoing hormone therapy for prostate cancer frequently experience a marked loss in their ability to grow facial hair. In addition to declines in libido and erectile function that often accompany such loss, men can also experience weight gain and enlarged breasts—a biological 'feminization' that can devastate a sense of the masculine self. It is not difficult to see how a celebration of facial hair—chosen, after all, for its connotations of virility and potency—might be hard to bear for men undergoing treatment.

Leaving aside the issues of messaging and metaphors, and the claims that men somehow represent a medically marginalized group, I'd like to move on now to consider in more detail the basis of the claim that prostate cancer is a 'neglected' disease.

A 'Neglected' Disease?

To a greater or lesser extent (depending on the organization), the rhetoric of exclusion and neglect has been a part of the identity and rationale for prostate activism ever since these groups first began to formally organize in the 1980s. There are some very good reasons why this should be the case. Claims to exclusion and neglect are inherently provocative, implying an essential unfairness about the way the world works. Activists, by their nature, tend to rely on provocation of some sort—whether to thought, to anger, or to action, or to all three—as the means to move against apathy and complacency and to seed attention-grabbing conversation and debate. Provocative language serves its purpose. There is another good reason why prostate cancer activists sometimes present cancer research as kind of zero-sum game: taken at face value the data appear to very much support such a view. Even a cursory review of figures released in the past few decades by federal and major charitable bodies will show that they have *not* spent their money in evenhanded ways. Some cancers simply *do* receive more research funding than others. Observers anticipating a distribution of research funding that reflects, say, the overall numbers of people affected by a particular cancer, or some particular commitment to cancers for which there are few treatments and little hope, might well be shocked at what might seem like an unconscionable injustice at work. [17] So what *is* going on?

A disease like lung cancer might lack the visibility and research support of a disease like childhood leukaemia, but the reasons why, as the historian Carsten Timmermann argues, can be deceptively complex.[18] Social factors likely play a role in the discrepancy, especially since smokers who develop lung cancer are routinely stigmatized as the 'deserving' sufferer, a disease identity in obvious contrast to that of the innocent child dealt a cruel and random blow. As Timmermann says, though, such factors are only part of the story (and perhaps a small part at that). Of greater significance is the fact that diseases that are clinically intractable tend to attract less research attention because they also tend to be less well understood at the basic, biological level. It makes sense that this should be so. The kinds of pathophysiological insights that come from therapeutic interventions in a disease not only fuel further investigation of how that disease might be optimally managed, they also tell us a lot about how the disease itself functions. Sometimes, as Timmermann shows, a recalcitrant and low profile disease like lung cancer can experience a quite sudden a spike in research funding and interest when the questions that would be asked of a disease co-occur with some attainable means of answering them. If hopes then fade for a breakthrough, as it did in lung cancer research when a brief period of optimism was followed by a period of pessimism in the 1970s, funding will likely shrink again. Funding and interest are not linked to a disease so much as they are linked to the questions that the disease might help answer.

Charles Huggins' work on the hormone-dependency of prostate cancer described in Chap. 4 is a good example of how tractability attracts interest. His discovery that prostatic carcinomas could and did spontaneously arise in elderly dogs was transformative for him. By this I mean not just that the animal model provided the practical experimental means by which Huggins was able to pry open and observe the relationship between circulating hormones and the physiological life of a carcinoma.

In the absence of the laboratory model, it would seem profoundly unlikely that an ambitious young investigator like Huggins—one who had the run of the extensive and expensive new facilities built by his employer, the University of Chicago, in the late 1920s to better frame their medical school in the terms of the new 'academic medicine' then coming into vogue—would have saddled himself with an intellectually sterile disease like prostate cancer. It is, in other words, a feature and a consequence of the success of biomedicine that it seeks out puzzles that best lend themselves to the methods of solving most prized by its investigators.

Similarly, as I discuss in Chap. 5, the reason why acute lymphocytic leukaemia in children became such a focus for the post-WWII NCI was that it seemed to offer the kinds of questions that might plausibly be solved within a short period of time. It, too, was a cancer that could be studied in detail in the laboratory thanks to the availability of a mouse model. The chemotherapies that emerged from these investigations were in their turn well suited to the application of the new tool of clinical investigation, the RCT. The information produced by these trials provided invaluable data about the optimal drug combinations and dosage and, in doing so, laid bare some of the essential biological nature of the cancer itself. It was a potent combination for the investigators who were then able to use the knowledge gained to effect dramatic improvements to the survival rates. It was also a huge vindication of the research strategies nurtured by the NCI, and brought a good deal of prestige to the institution and to biomedicine more generally. This is not to say that curiosity trumped compassion. As Emm Barnes Johnstone and Joanna Baines make clear in their history of childhood cancer, NCI Clinical Center (and later MDACC) researchers like the two Emils—Frei and Freireich—could be flawlessly dedicated and creative clinicians, even as they were simultaneously highly ambitious academics.[19] As other paediatric cancers proved to be malleable to comparable methods of laboratory and clinical investigation in the laboratory (with tumour lines later joining animal models), so research questions were expanded and interest sustained. Today the majority of all pediatric patients diagnosed with cancer (in the industrialized world) are enrolled in at least one clinical trial—some are for the testing of new drugs or drug combinations, more still look for ways to reduce the toxic effects and long-term consequences of treatment. Again, this ongoing cascade of questions and (solvable) problems helps to maintain paediatric cancers firmly in the public eye and, indeed, the public purse. With the recent improvements in mouse models for cancers of the prostate, lung, and pancreas, it remains to be seen if the 'neglected' cancers will see a surge of interest and funding. Historical indicators would suggest that this will be so.

The availability of nonhuman experimental tools (like tissue cultures or animal models) can also lead to other, less obvious, consequences for levels of funding. As I discuss in Chap. 7, part of the reason why Ernest Lawrence wrapped his cyclotron around specifically *medical* questions was to find funding for research that he might otherwise not have been able to carry out (in his case he achieved this by appealing to the medicine and public health emphasis in the philanthropy of the Rockefeller Foundation). An industry of basic science and basic scientists followed his lead. One of my favourite medical thinkers of the twentieth century, Lewis Thomas, stressed that problems of the 'halfway' technologies (that is, technologies like dialysis that supported life without eliminating disease) that were arising with some considerable frequency in the latter third of the twentieth century required us to refocus and return to the laboratory to ask questions about basic disease mechanisms.[20] The somewhat tenuous connection between real life clinical problems and much of what is regarded (and funded) today as basic 'medical' research, is surely not what Lewis had in mind for us, but, in any case such spending is included in the statistics for diseases amenable to laboratory research. This money, in turn, can (and obviously does) give the impression that the eradication of some diseases is 'valued' over efforts to control others, and so that some patients, some people, are valued over others. The reality is much more complex than that.

This book might then be considered to be (at least tangentially) a part of the project that Timmermann lays out for himself in his book—the writing of history in the absence of progress. My book might at first glance appear to be no such thing. I am after all dealing with prostate cancer, the subject of some of the most dramatic claims to progress made by late twentieth century medicine. Accounts of 'progress' in prostate cancer are almost always about the 'cure' of a disease caught early, and while this claim is in itself problematic, it also makes us think about what is left out—those 'recurring' and late stage cancers for which the prognoses remain dismal. Recalcitrance has, in this instance, helped engineer an obsession with early detection that has been so effective in its reach that few American men reach middle age without feeling it.

QUESTIONS AND CONCERNS FOR THE FUTURE

When the Johns Hopkins urologist Donald Coffey asserted at the 1992 annual meeting of the American Cancer Society that the prostate was 'the major site of medical problems in the American man', it was no hyperbole.[21] The millennia-old problem of urinary strangulation caused by benign prostatic enlargement was by the early 1990s one of the major reimbursement costs for Medicare, as treatments and surgeries multiplied to meet the needs of America's aging baby-boomers. With increases in longevity and decreases in smoking rates, prostate malignancy had by then replaced lung cancer as the most commonly diagnosed cancer in the US, and remained only second to it as the leading cause of cancer deaths for the American male. These issues and prostate disease itself were not just issues for the *American* man, of course—the world over, prostate cancer is one of the most commonly diagnosed cancers in men. In some parts of the world though, notably China and Japan, cancer of the prostate (as, indeed, its female analogue, cancer of the breast) is much more rarely seen. Moreover, when migration occurs from countries of low incidence to those where prostate cancer occurs with greater frequency, cancer rates increase in the migrants but not to the levels seen in Europeans and white Americans—observations that suggest there is some genetic component to the disease in addition to environmental and lifestyle factors.[22] Men of African descent, on the other hand, show some of the highest rates of incidence of all, something prostate cancer researchers in the US realized decades ago.

In addition to their RCT studies of prostate cancer therapies, Gerald Murphy and the National Prostatic Cancer Project (NPCP) also participated in major epidemiological surveys to understand the distribution of risk across the population. As they expected, older men were more at risk of developing prostate cancer than younger men, but the more startling finding was to do with the high rates of disease amongst African-American men. Murphy's results also suggested that these men, once affected, were significantly less likely to survive than their white counterparts.[23] The literature on the history and present circumstances of health disparities in US healthcare is large and growing. The historian Keith Wailoo describes how when the PSA test became available some of it supporters 'waved the red flag of racial disparities—insisting that the test might help reduce those disparities, ... [claims that were] a mixture of hope and marketing bravado'.[24] Some researchers suggested tweaking the test by using different scales when applied to different racial groups. For others, though, this overemphasis on biological factors marginalized what they perceived to be the other more significant (and significantly more complex) problems of health disparities, like economic and social status, health insurance coverage status, and so on. For African-American men put on the pathway to biopsy and surgery, outcomes were poorer than for other groups, and this we might (unfortunately) expect, given what the data tell us about health disparities.[25] Affluent men and (in later chapters) men with good private health insurance or those older men able to cover the numerous gaps in Medicare coverage have been (with the exception perhaps of the VA story) the implicit focus of this book; the stories of those poorer more socially marginal men who tend to have more piecemeal access to—and a fragmented experience of—US healthcare are largely still to be told.

So what is the future for men diagnosed for prostate cancer in the United States? There is, of course, an enormous amount left out of this book. Aside from a brief mention of finasteride, I have not, for instance, talked about the post-DES story of hormone ablation therapy. Neither have I included the surgical advances, like nerve-sparing operations, that were developed in the latter half of the twentieth century. The treatment of men with advanced cancer also receives scant consideration in this book. Innovations in chemotherapy (to treat bone metastases in particular) and in the clinical and psychosocial support such men receive from the medical and patient advocacy communities have changed the lives of many living with prostate cancer, as it has their families. I should know: I belong to a family like that. I am the daughter of a father with metastatic prostate cancer and it is the experiences of this remarkable man as he has lived with his disease that inspired me to write this book. I have had the opportunity to see up close how his, and by extension our, lives have been eased and improved thanks to new trends and treatments. Supporting a father cared for in a very different healthcare system than the one with which I am most familiar, has, however, given me pause for thought.

My father's cancer was first diagnosed in the late 2000s because of the pain resulting from the bony metastases then invading his spine, seeded by a primary tumour growing in and spreading from his prostate. His was a situation not too uncommon an occurrence for men who, like my father, live in France. For a man living here in the United States of comparable age (early sixties), class (middle), and insurance status (excellent), such occurrences are by contrast vanishingly rare. In the anger I felt following his diagnosis, I raged at a system that I believed had allowed him to become needlessly sick. Why had his PSA levels been allowed to rise so high before they were even checked? Why hadn't he been urged to attend regular screenings? Wouldn't these things have saved him from the cancer disseminating his body? The answer, as I learned after reading a lot and bending the ears of some insightful (and patient) physician-colleagues was that it might have; then again it might not have made much of a difference at all.

Cancer biology is a tricky thing. Regardless of tumour site, what may appear at first diagnosis to be a very treatable, early-stage cancer can sometimes nevertheless possess an aggressiveness that rapidly overwhelms and kills the patient. Similarly, metastatic cancers with their typically poor prognosis can sometimes be eminently susceptible to treatments that cause them to shrink and stabilize, allowing the patients they affect years of good health. This latter scenario did, mercifully, prove to be the case for my father's primary tumour. If he had been diagnosed much earlier, as a result of a PSA test say, he might now be cancer free, or he might be a man as he is now who lives with prostate cancer. If he were an American, my father would almost certainly have spent years by now without his prostate, dealing with the various issues attendant upon that and the after-effects of the radiotherapy he would have received, but even so he might still have cancer.

Despite the massive push to heroic intervention, deaths from prostate cancer in the US remain high, second only, as I have referred to several times in this book, to lung cancer in cancer deaths in men. What the US system *does* have on its side are rates of survival—the French rates look rather miserable in comparison—but as I discussed in Chap. 6, pushing the clock back on diagnosis can produce misleading survival data. Take a hypothetical case in which two seventy-five year old men, one American, one French, die of prostate cancer. The American man may well have been diagnosed and aggressively treated in his mid-fifties, dying following a 'recurrence' of his disease. The Frenchman, on the other hand, might well have been diagnosed at seventy after he (like my father) complained to his general practitioner of back pain caused by metastatic cancer. Cancer statistics based on cases like these would show the American as surviving fifteen years longer than his French counterpart. For this reason, survival statistics taken at face value are a poor measure of progress, and they are all too frequently used to support the philosophy of intervention and to sway the unwary reader—like me back in the day when I was looking for someone to take responsibility for my father having cancer. This is not to say that French physicians are lazy in their treatment of cancer. In my direct experience, nothing could be further from the case. When high technology is called for, they use it: my father has enjoyed world-class care and the latest therapies. It is, however, pressure at a point, not the widespread 'more is better' approach to technology, medication, and intervention that we use in the US (sometimes to excellent effect: in some important ways the US *does* have the best healthcare in the world, that is why the Texas Medical System and others like it attract referrals from across the globe). We can and do make ourselves sick with overtreatment and overdiagnosis and as a country we pay the many physical, emotional, and financial costs of doing so.

After all this research done and knowledge gained, I still sometimes think about whether my father might have faired better if he lived with me here in the US. My answer to that is, possibly. I know for sure though that for the sake of *his* cancer being eradicated (maybe, maybe not) for good, many, many more men in addition would have been needlessly exposed to interventions that are costly, anxiety provoking, potentially disabling, and possibly dangerous. Some might see this as a necessary, if painful, trade-off, but I, on balance, do not. While I will always, I'm sure, wonder what might have been, I am ultimately glad that my father is as he is tucked away in a French village experiencing all the fine expertise and support that his physicians and nurses provide.

Much of this book—especially in the latter chapters—is focused on the problems, on the scandals, and on the controversies that surrounded prostate cancer in the United States. It's a focus that easily lends itself to accusations of 'doctor-bashing' on my part. Participants in the field of the history and philosophy of science, technology, and medicine (or social studies of the same) are no strangers to accusations of anti-science bias. Given the kinds of questions we're interested in asking, and particularly the kinds of ways in which our accounts differ in tone and content from, for instance, many practitioner-penned histories, that's not too surprising. I am at my core a teacher and one who has spent the last ten years of her life teaching history to pre-professional and professional students (rather than, say, teaching history to history majors) as part of a large medical humanities programme within a city that is home to the world's largest medical centre, the Texas Medical Center. I do this job because I am inspired by my energetic and idealistic premeds, and I believe history (and the wider humanities) can bring them deep insights into the wider causes, context, and consequences of the actions they themselves will be responsible for as physicians. While I would not of course insist that my students agree with me (something that would be not only pointless but also hopeless: these students tend to be smart and curious, in parts committed and contrary), I do ask them to see the world as complicated and to understand how and why their passionate intentions to do good can sometimes turn treacherous.

The unintended, and problematic, consequences that we sometimes unwittingly unleash by our decisions and actions can take us all by surprise, whoever we are. That trying to do good can, with some regularity, lead to poor outcomes, may strike us as baffling and frustrating, even alienating: our most earnest good intentions spited. If we take 'doctor' to still signify something of its Latin root *docere*—to teach—then these future teachers would do well to bear these burdens of complexity to better guide and support patients and their families who might be inclined to see issues in more black and white terms. For my students oozing enthusiasm and idealism for their future practice, I offer them questions and arguments like the ones I present in this book not to accuse them or berate them but rather to, in whatever small way, to help them on their way to honing the skills of compassion and empathy that they themselves admire and strive to attain. These questions do not then, I think, represent the principles of doctor-bashing so much as they touch on (or try to, at least) some of the most foundational issues in a fearless, humanistic education.

Notes

1. Coffey, Prostate Cancer, 880.
2. Aronowitz, *Unnatural History*.
3. Löwy, *A Woman's Disease*.
4. Rosenberg, Managed Fear.
5. Aronowitz, *Unnatural History*, 419.
6. Vos, *Drugs Looking for Diseases*.
7. Miele and Clarke, We Remain Very Much the Second Sex.
8. Clarke, A Comparison of Breast, Testicular and Prostate Cancer, 548.
9. My thanks to Carsten Timmermann for pointing out this age bias in anti-breast cancer campaigns.
10. Chen et al., Twenty Years Post-NIH Revitalization Act.
11. Goldberg, *Women Are Not Small Men*.
12. American College of Cardiology, Women Don't Get to Hospital Fast Enough during Heart Attack.
13. Wenner, The War against War Metaphors.

14. Wassersug, Oliffe, and Han, On Manhood and Movember.
15. Ibid., 65.
16. Movember, Movember United States.
17. Parker-Pope, Cancer Funding.
18. Timmermann, *A History of Lung Cancer*, 175.
19. Johnstone and Baines, *The Changing Faces of Childhood Cancer*.
20. Thomas, On the Science and Technology of Medicine.
21. Coffey, Prostate Cancer, 880.
22. James, *Cancer: A Short Introduction*, 4.
23. Mettlin and Murphy, Cancer among Black Populations.
24. Wailoo, *How Cancer Crossed the Color Line*, 154.
25. Ibid., 159–60.

BIBLIOGRAPHY

Aaronson, D.S., A.Y. Odisho, and N. Hills. 2012. Proton Beam Therapy and Treatment for Localized Prostate Cancer: If You Build It, They Will Come. *Archives of Internal Medicine* 172: 280–283.

Abel, Gregory, Richard Penson, Steven Joffe, Lidia Schapira, Bruce Chabner, and Thomas Lynch. 2006. Direct-to-Consumer Advertising in Oncology. *The Oncologist* 11: 217–226.

Abernethy, John. 1814. *The Surgical Works of John Abernethy*. London: Longman, Hurst, Rees, Orme, and Brown.

Ablin, Richard, and Ronald Piana. 2014. *The Great Prostate Hoax: How Big Medicine Hijacked the PSA Test and Caused a Public Health Disaster*. New York, NY: Palgrave Macmillan.

Ablin, Richard. n.d. The Great Prostate Mistake. *The New York Times*, March 9, 2010.

Ackerknecht, Erwin. 1967. *Medicine at the Paris Hospital, 1794–1848*. Baltimore: Johns Hopkins Press.

Adams, John. 1853. The Case of Scirrhus of the Prostate Gland, with a Corresponding Affection of the Lymphatic Glands in the Lumbar Region and in the Pelvis. *Lancet* 61: 393–394.

AHRQ. n.d. Particle Beam Radiation Therapies for Cancer—Executive Summary AHRQ Effective Health Care Program. September 14, 2009. Accessed September 26, 2015. http://effectivehealthcare.ahrq.gov/index.cfm/search-for-guides-reviews-and-reports/?pageaction=displayproduct&productid=174

Albertsen, Peter, James Hanley, George Barrows, David Penson, Pam Kowalczyk, Melinda Sanders, and Judith Fine. 2005. Prostate Cancer and the Will Rogers Phenomenon. *Journal of the National Cancer Institute* 97: 1248–1253.

© The Editor(s) (if applicable) and The Author(s) 2016   209
H. Valier, *A History of Prostate Cancer*,
DOI 10.1057/978-1-137-56595-2

American College of Cardiology. n.d. Women Don't Get to Hospital Fast Enough During Heart Attack. *Heart Disease Weekly*, March 29, 2015. Accessed July 31, 2015 ScienceDaily, 5 March 2015. http://www.sciencedaily.com/releases/2015/03/150305205955.htm

Aminoff, Michael. 2011. *Brown-Séquard an Improbable Genius Who Transformed Medicine*. New York: Oxford University Press.

Andrews, G.S. 1949. Latent Carcinoma of the Prostate. *Journal of Clinical Pathology* 2: 197–208.

Andriole, Gerald, David Crawford, Robert Grubb, Saundra Buys, David Chia, Timothy Church, et al. 2009. Mortality Results from a Randomized Prostate-Cancer Screening Trial. *The New England Journal of Medicine* 360: 1310–1319.

Arnst, Catherine. n.d. A Gender Gap in Cancer. *BloombergView*, June 13, 2007. Accessed September 3, 2015. http://www.bloomberg.com/bw/stories/2007-06-13/a-gender-gap-in-cancerbusinessweek-business-news-stock-market-and-financial-advice

Aronowitz, Jesse. 2012. Whitmore, Henschke, and Hilaris: The Reorientation of Prostate Brachytherapy (1970–1987). *Brachytherapy* 11: 157–162.

Aronowitz, Robert. 2014. From Skid Row to Main Street: The Bowery Series and the Transformation of Prostate Cancer, 1951–1966. *Bulletin of the History of Medicine* 88: 287–318.

———. 2014. "Screening" for Prostate Cancer in New York's Skid Row: History and Implications. *American Journal of Public Health* 104: 70–76.

———. 2007. *Unnatural History: Breast Cancer and American Society*. Cambridge: Cambridge University Press.

———. 2001. Do Not Delay: Breast Cancer and Time, 1900–1970. *The Milbank Quarterly* 79: 355–386.

American Urological Association. n.d. Detection of Prostate Cancer. Accessed September 15, 2015. https://www.auanet.org/education/guidelines/prostate-cancer-detection.cfm

———. n.d. AUA Responds to New Recommendations on Prostate Cancer Screening. Accessed September 14, 2015. https://www.auanet.org/advnews/press_releases/article.cfm?articleNo=262

———. n.d. AUA Applauds Passage of New Jersey Legislation Opposing USPSTF PSA Testing Recommendations and Guaranteeing Coverage. Accessed September 14, 2015. http://www.auanet.org/advnews/press_releases/article.cfm?articleNo=266

Bagshaw, Malcolm, Henry Kaplan, and Robert Sagerman. 1965. Linear Accelerator Supervoltage Radiotherapy. *Radiology* 85: 121–129.

Bailar, John. 1997. Editorial: The Promise and Problems of Meta-Analysis. *The New England Journal of Medicine* 337: 559–561.

Baker, Robert. 2013. *Before Bioethics: A History of American Medical Ethics from the Colonial Period to the Bioethics Revolution.* Oxford: Oxford University Press.

Barclay, Laurie. n.d. End of an Era for PSA: A Newsmaker Interview With Thomas Stamey, MD. September 17, 2004. Medscape. Accessed September 15, 2015. http://www.medscape.com/viewarticle/489474

Barr, Donald. 2011. Revolution or Evolution? Putting the Flexner Report in Context. *Medical Education* 45: 17–22.

Barringer, Benjamin. 1931. Carcinoma of the Prostate. *Annals of Surgery* 93: 326–335.

Bentzen, Søren. 2008. Radiation Oncology Health Technology Assessment: The Best Is the Enemy of the Good. *Nature Clinical Practice Oncology* 5: 563.

Bernard, Claude. 2012. *An Introduction to the Study of Experimental Medicine.* New York: Courier Corporation.

Billroth, Theodore. 1872. General Surgical Pathology and Therapeutics, in Fifty-One Lectures. *Internet Archive.* Accessed July 23, 2015. https://archive.org/details/generalsurgicalp1879bill

———. 1869. Chirurgische Klinik, Zürich, 1860–1867. *Internet Archive.* Accessed July 23, 2015. https://archive.org/details/chirurgischekli03billgoog

Binney, Horace. 1914. Cancer of the Prostate. *Boston Medical and Surgical Journal* 171: 748–752.

Bloom, David, and Frank Hinman Jr. 2003. Frank Hinman, Sr: A First Generation Urologist. *Urology* 61: 876–881.

Blue, Ethan. 2009. The Strange Career of Leo Stanley: Remaking Manhood and Medicine at San Quentin State Penitentiary, 1913–1951. *Pacific Historical Review* 78: 210–241.

Boston Women's Health Book Collective, ed. 1976. *Our Bodies Ourselves. A Book by and for Women. The Boston Women's Health Book Collective. 2nd. ed., Completely Rev. and Expanded.* New York: Simon, Schuster.

Brada, Michael, Madelon Pijls-Johannesma, and Dirk De Ruysscher. 2009. Current Clinical Evidence for Proton Therapy. *Cancer Journal* 15: 319–324.

Brada, Michael. 2007. Proton Therapy in Clinical Practice: Current Clinical Evidence. *Journal of Clinical Oncology* 25: 965–970.

Brieger, Gert. 1992. From Conservative to Radical Surgery in Late Nineteenth Century America. In *Medical Theory Surgical Practice: Studies in the History of Surgery*, ed. Christopher Lawrence, 216–231. New York: Routledge.

Brown-Séquard, Charles. 1889. Note on the Effect Produced on Man by Subcutaneous Injections of a Liquid Obtained from the Testicles of Animals. *Lancet* 134: 105–107.

Bucci, Kara, Alison Bevan, and Mack Roach. 2005. Advances in Radiation Therapy: Conventional to 3D, to IMRT, to 4D, and Beyond. *CA: A Cancer Journal for Clinicians* 55: 117–134.

Bud, Robert. 1978. Strategy in American Cancer Research After World War II. *Social Studies of Science* 8: 425–459.

Bumpus, Hermon. 1926. Carcinoma of the Prostate. *Journal of Surgery, Gynecology & Obstetrics* 43: 150–155.

Burns, L.R., and J.C. Robinson. 1997. Physician Practice Management Companies: Implications for Hospital-Based Integrated Delivery Systems. *Frontiers of Health Services Management* 14: 3–35.

Bynum, William. 1994. *Science and the Practice of Medicine in the Nineteenth Century*. Cambridge: Cambridge University Press.

Cabot, Arthur Tracy. 1896. II. The Question of Castration for Enlarged Prostate. *Annals of Surgery* 24: 265–309.

Cabot, Hugh. 1912. Is Urology Entitled to Be Regarded as a Specialty? *Transactions of the American Urological Association* 5: 1–20.

Carlsson, Sigrid, Andrew Vickers, Hans Lilja, and Jonas Hugosson. 2012. Screening for Prostate Cancer. *Annals of Internal Medicine* 156: 539.

Catalona, William J., Anthony V. D'Amico, William F. Fitzgibbons, Omofolasade Kosoko-Lasaki, Stephen W. Leslie, Henry T. Lynch, Judd W. Moul, Marc S. Rendell, and Patrick C. Walsh. 2012. What the U.S. Preventive Services Task Force Missed in Its Prostate Cancer Screening Recommendation. *Annals of Internal Medicine* 157: 137–138.

Catalona, William. 1991. Measurement of Prostate-Specific Antigen in Serum as a Screening Test for Prostate Cancer. *New England Journal of Medicine* 324: 1156–1161.

Chalmers, Iain, and Mike Clarke. 2004. Commentary: The 1944 Patulin Trial: The First Properly Controlled Multicentre Trial Conducted under the Aegis of the British Medical Research Council. *International Journal of Epidemiology* 33: 253–260.

Chen, Moon, Primo Lara, Julie Dang, Debora Paterniti, and Karen Kelly. 2014. Twenty Years Post-NIH Revitalization Act: Enhancing Minority Participation in Clinical Trials (EMPaCT): Laying the Groundwork for Improving Minority Clinical Trial Accrual: Renewing the Case for Enhancing Minority Participation in Cancer Clinical Trials. *Cancer* 120: 1091–1096.

Chou, R., J.M. Croswell, T. Dana, C. Bougatsos, I. Blazina, R. Fu, K. Gleitsmann, et al. 2011. Screening for Prostate Cancer: A Review of the Evidence for the U.S. Preventive Services Task Force. *Annals of Internal Medicine* 155: 762–771.

Clarke, Juanne. 2004. A Comparison of Breast, Testicular and Prostate Cancer in Mass Print Media (1996–2001). *Social Science & Medicine* 59: 541–551.

Coffey, Donald. 1993. Prostate Cancer. An Overview of an Increasing Dilemma. *Cancer* 71: 880–886.

Collins, M.M., F.J. Fowler Jr., R.G. Roberts, J.E. Oesterling, G.J. Annas, and M.J. Barry. 1997. Medical Malpractice Implications of PSA Testing for Early

Detection of Prostate Cancer. *The Journal of Law, Medicine & Ethics* 25: 234–242.

Cooper, J. 1965. Onward the Management of Science: The Wooldridge Report: NIH Was Not Cleared on All Counts by the Wooldridge Committee, Which Itself Is Rated Low on Methodology. *Science* 148: 1433–1439.

Cooper, Astley, and Frederick Tyrrell. 1831. *The Lectures of Sir Astley Cooper: On the Principles and Practice of Surgery*. London: Lilly and Wait.

Cotter, Dennis. n.d. The National Center For Health Care Technology: Lessons Learned. *Health Affairs* January 22, 2009. Accessed September 27, 2015. http://healthaffairs.org/blog/2009/01/22/the-national-center-for-health-care-technology-lessons-learned/

Cox, James. 2008. Assessment of Technological Advances in Radiation Oncology: From Whence Comes the Funding? *International Journal of Radiation Oncology, Biology, Physics* 72: 11.

———. 1995. Evolution and Accomplishments of the Radiation Therapy Oncology Group. *International Journal of Radiation Oncology, Biology, Physics* 33: 747–754.

Cunningham, Andrew. 2010. *The Anatomist Anatomis'd: An Experimental Discipline in Enlightenment Europe*. Farnham: Ashgate Publishing.

Daly, Jeanne. 2005. *Evidence-Based Medicine and the Search for a Science of Clinical Care*. Berkeley: University of California Press.

David, K., E. Dingemanse, J. Freud, and E. Lacqueur. 1935. Über Krystallinisches Männliches Hormon Aus Hoden (Testosteron), Wirksamer Als Aus Harn Oder Aus Cholesterin Bereitetes Androsteron. *Hoppe-Seyler's Zeitschrift Für Physiologische, Chemie* 233: 281.

Davis, Rebecca L. 2008. "Not Marriage at All, but Simple Harlotry": The Companionate Marriage Controversy. *The Journal of American History* 94: 1137–1163.

DeAntoni, E.P. 1997. Eight Years of "Prostate Cancer Awareness Week": Lessons in Screening and Early Detection. Prostate Cancer Education Council. *Cancer* 80: 1845–1851.

Deaver, John. 1905. *Enlargement of the Prostate: Its History, Anatomy, Etiology, Pathology, Clinical Causes, Symptoms, Diagnosis, Prognosis, Treatment; Technique of Operations, and After-Treatment*. Boston: P. Blakiston, Son & Company.

Delahunt, Brett, R.J. Miller, J.R. Srigley, A.J. Evans, and H. Samaratunga. 2012. Gleason Grading: Past, Present and Future. *Histopathology* 60: 75–86.

Desault, Pierre Joseph. 1791. *Traité des Maladies des Voies Urinaires*. Ghent: Chez l'Auteur.

DeVita, Vincent. 1983. The Governance of Science at the National Cancer Institute: A Perspective on Misperceptions. *Cancer Research* 43: 3969–3973.

Donohue, Julie. 2006. A History of Drug Advertising: The Evolving Roles of Consumers and Consumer Protection. *The Milbank Quarterly* 84: 659–699.

Drago, J.R. 1989. The Role of New Modalities in the Early Detection and Diagnosis of Prostate Cancer. *CA: A Cancer Journal for Clinicians* 39: 326–336.

Drazer, Michael, Dezheng Huo, and Scott Eggener. 2015. National Prostate Cancer Screening Rates After the 2012 US Preventive Services Task Force Recommendation Discouraging Prostate-Specific Antigen-Based Screening. *Journal of Clinical Oncology* 33: 2416–2423.

Du Laurens, André. 1600. *Historia Anatomica Humani Corporis*. Paris: Marcus Orry.

Duncan, W., S.J. Arnott, J.J. Battermann, J.A. Orr, G. Schmitt, and G.R. Kerr. 1984. Neutrons in the Treatment of Head and Neck Cancers: The Results of a Multi-Centre Randomly Controlled Trial. *Radiotherapy and Oncology* 4: 293–300.

Eastman, Joseph. 1909. Confessions of a Yeoman Prostatectomist. *Transactions of the American Urological Association* 2: 142–152.

Efstathiou, Jason. n.d. 'Proton Therapy vs. IMRT for Low or Intermediate Risk Prostate Cancer' ClinicalTrials.gov. Accessed September 26, 2015. https://clinicaltrials.gov/ct2/show/NCT01617161?term=PARTIQoL&rank=1#wrapper

Editorial. 1939. Endocrinology of the Prostate. *Lancet* 233: 1339–1341.

Endicott, Kenneth. 1957. The Chemotherapy Program. *Journal of the National Cancer Institute* 19: 275.

Epstein, K. 2012. Is Spending on Proton Beam Therapy for Cancer Going Too Far, Too Fast? *British Medical Journal* 344: e2488.

Evans, Imogen, ed. 2011. *Testing Treatments: Better Research for Better Healthcare*. London: Pinter & Martin.

Evidence-Based Medicine Working Group. 1994. Evidence-Based Health Care: A New Approach to Teaching the Practice of Health Care. Evidence-Based Medicine Working Group. *Journal of Dental Education* 58: 648–653.

Farber, Sidney, Louis Diamond, Robert Mercer, Robert Sylvester, and James Wolff. 1948. Temporary Remissions in Acute Leukemia in Children Produced by Folic Acid Antagonist, 4-Aminopteroyl-Glutamic Acid (Aminopterin). *New England Journal of Medicine* 238: 787–793.

Feinstein, A.R., D.M. Sosin, and C.K. Wells. 1985. The Will Rogers Phenomenon—Stage Migration and New Diagnostic Techniques as a Source of Misleading Statistics for Survival in Cancer. *New England Journal of Medicine* 25: 1604–1608.

Feldstein, Dan. n.d. Police, Fire Pensions Warned about Deal. *Houston Chronicle*, October 23, 2005. Accessed September 7, 2015. http://www.chron.com/news/houston-texas/article/Police-fire-pensions-warned-about-deal-1487556.php

————. n.d. Proton-Therapy Costs vs. Benefits Debated. *Houston Chronicle.* October 23, 2005. Accessed September 7, 2015. http://www.chron.com/news/houston-texas/article/Proton-therapy-costs-vs-benefits-debated-1920064.php

Fintor, Lou. 1999. For-Profit Treatment Centers: Trailblazing a New Model of Care? *Journal of the National Cancer Institute* 91: 1272–1274.

————. 2002. Direct-to-Consumer Marketing: How Has It Fared? *Journal of the National Cancer Institute* 94: 329–331.

Fowler, J.F. 1984. 40 Years of Radiobiology: Its Impact on Radiotherapy. *Physics in Medicine and Biology* 29: 97–113.

Freedman, B. 1987. Equipoise and the Ethics of Clinical Research. *The New England Journal of Medicine* 317: 141–145.

Freeman, E., D. Bloom, and E. McGuire. 2001. A Brief History of Testosterone. *The Journal of Urology* 165: 371–373.

Frei, Emil, James F. Holland, Marvin A. Schneiderman, Donald Pinkel, George Selkirk, Emil J. Freireich, Richard T. Silver, G. Lennard Gold, and William Regelson. 1958. A Comparative Study of Two Regimens of Combination Chemotherapy in Acute Leukemia. *Blood* 13: 1126–1148.

Gardner, Faxton. 1907. Some Remarks on Prostatic Carcinoma. *The American Journal of Urology* 3: 394–408.

Gaudillière, Jean-Paul. 2005. Better Prepared than Synthesized: Adolf Butenandt, Schering Ag and the Transformation of Sex Steroids into Drugs (1930–1946). *Studies in History and Philosophy of Biological and Biomedical Sciences* 36: 612–644.

Gawande, Atul. n.d. The Cost Conundrum. *The New Yorker*, June 1, 2009. Accessed September 27, 2015. http://www.newyorker.com/magazine/2009/06/01/the-cost-conundrum

Gawande, Atul. 2002. *Complications: A Surgeon's Notes on an Imperfect Science.* New York: Metropolitan Books.

Gelband, Helen. 1983. *The Impact of Randomized Clinical Trials on Health Policy and Medical Practice: Background Paper.* Office of Technology Assessment: Congress of the United States.

Gigerenzer, G., J. Mata, and R. Frank. 2009. Public Knowledge of Benefits of Breast and Prostate Cancer Screening in Europe. *Journal of the National Cancer Institute* 10: 1216–1220.

Gijswijt-Hofstra, Marijke, and Roy Porter. 2001. *Cultures of Neurasthenia from Beard to the First World War.* Clio Medica 63. Amsterdam: Rodopi.

Gilman, A. 1946. Therapeutic Applications of Chemical Warfare Agents. *Federation Proceedings* 5: 285–292.

Gleason, Donald. 1992. Histologic Grading of Prostate Cancer: A Perspective. *Human Pathology* 23: 273–279.

————. 1966. Classification of Prostatic Carcinomas. *Cancer Chemotherapy Reports Part* 50: 125–128.

Gleason, Donald, and G.T. Mellinger. 1974. Prediction of Prognosis for Prostatic Adenocarcinoma by Combined Histological Grading and Clinical Staging. *The Journal of Urology* 111: 58–64.

Goitein, Michael, and James Cox. 2008. Should Randomized Clinical Trials Be Required for Proton Radiotherapy? *Journal of Clinical Oncology* 26: 175–176.

Goldberg, Nieca. 2002. *Women Are Not Small Men: Life-Saving Strategies for Preventing and Healing Heart Disease in Women*. New York: Ballantine Books.

Gooday, Graeme. 2008. Placing or Replacing the Laboratory in the History of Science? *Isis* 99: 783–795.

Goodman, Jordan, Anthony McElligott, and Lara Marks. 2003. *Useful Bodies: Humans in the Service of Medical Science in the Twentieth Century*. Baltimore: Johns Hopkins University Press.

Gouley, John. 1885. Some Points in the Surgery of the Hypertrophied Prostate. *Transaction of the American Surgical Association* 3: 163–192.

Government Accountability Office. n.d. Medicare: Higher Use of Costly Prostate Cancer Treatment by Providers Who Self-Refer Warrants Scrutiny. July 19, 2013. Accessed September 27, 2015. http://www.gao.gov/products/GAO-13-525

Grabstald, H. 1965. Biopsy Techniques in the Diagnosis of Cancer of the Prostate. *CA: A Cancer Journal for Clinicians* 15: 134–137.

Griffin, Thomas W. 1992. Fast Neutron Radiation Therapy. *Critical Reviews in Oncology/Hematology* 13: 17–31.

Gronvall, J.A. 1989. The VA's Affiliation with Academic Medicine: An Emergency Post-War Strategy Becomes a Permanent Partnership. *Academic Medicine* 64: 61–66.

Gurll, N.J., J.W. Holcroft, P. Numann, D.G. Reynolds, R.S. Rhodes, R.P. Saik, F.T. Thomas, M.D. Tilson, and C.K. Zarins. 1982. The Veterans Administration and Academic Surgery. A Report from the Committee on Issues of the Association for Academic Surgery 1979 Meeting. *The Journal of Surgical Research* 33: 1–10.

Halperin, Edward C. 2006. Particle Therapy and Treatment of Cancer. *The Lancet Oncology* 7: 676–685.

Hamilton, David. 1986. *The Monkey Gland Affair*. London: Chatto & Windus.

Hansen, B. 1999. New Images of a New Medicine: Visual Evidence for the Widespread Popularity of Therapeutic Discoveries in America After 1885. *Bulletin of the History of Medicine* 73: 629–678.

Harvey, Dan. n.d. Proton Therapy. *Radiology Today Magazine*, September 7, 2009. Accessed September 7, 2015, http://www.radiologytoday.net/archive/090709p14.shtml

Heilbron, J.L., and Robert W. Seidel. 1989. *Lawrence and His Laboratory: A History of the Lawrence Berkeley Laboratory*. Berkeley: University of California Press.

Herbs, W. 1942. Biochemical Therapeusis in Carcinoma of the Prostate Gland: Preliminary Report. *Journal of the American Medical Association* 120: 1116–1122.

Hessel, Evan. n.d. Cancer Capitalists. *Forbes*, 11 November, 2006. Accessed September 27, 2015. http://www.forbes.com/forbes/2006/1127/178.html

Hey, William. 1805. *Practical Observations in Surgery, Illustrated with Cases and Plates*. Philadelphia: Humphries.

Hill, Catherine, and Agnès Laplanche. 2010. Prostate Cancer: The Evidence Weighs Against Screening. *Presse Médicale* 39: 859–864.

Hilts, Philip J. n.d. V.A. Hospital Is Told to Halt All Research. *The New York Times*, March 25, 1999. Accessed September 7, 2015, http://www.nytimes.com/1999/03/25/us/va-hospital-is-told-to-halt-all-research.html

Hippocrates, G.E.R., J. Lloyd, Chadwick, and W.N. Mann. 1983. *Hippocratic Writings*. Harmondswerth: Penguin Books.

Hitchcock, Alfred. 1842. Insanity and Death from Masturbation. *Boston Medical and Surgical Journal* 26: 283–286.

Hodges, Frederick. 2005. History of Sexual Medicine: The Antimasturbation Crusade in Antebellum American Medicine. *The Journal of Sexual Medicine* 2: 722–731.

Home, Everard. 1811. *Practical Observations on the Treatment of the Diseases of the Prostate Gland*. London: G. and W. Nicol.

Horan, Anthony H. 2012. *How to Avoid the Over-Diagnosis and Over-Treatment of Prostate Cancer*. Broomfield: On The Write Path Publishing.

Howe, Joseph. 1883. *Excessive Venery, Masturbation and Continence. The Etiology, Pathology and Treatment of the Diseases Resulting from Venereal Excesses, Masturbation and Continence*. New York: Bermingham & Company.

Hrushesky, W.J. 1994. The Department of Veterans Affairs' Unique Clinical Cancer Research Effort. *Cancer* 74: 2701–2709.

Huben, R.P., and Gerald Murphy. 1986. Prostate Cancer: An Update. *CA: A Cancer Journal for Clinicians* 36: 274–292.

Hudson, P.B., A.L. Finkle, A. Trifilio, and C.T. Wolan. 1954. Prostatic Cancer. IX. Value of Transurethral Biopsy in Search of Early Prostatic Carcinoma. *Surgery* 35: 897–900.

Huggins, Charles. 1966. Charles B. Huggins—Nobel Lecture: Endocrine-Induced Regression of Cancers. Accessed August 3, 2015. http://www.nobel-prize.org/nobel_prizes/medicine/laureates/1966/huggins-lecture.html

———. 1943. Endocrine Control of Prostatic Cancer. *Science* 97: 541–544.

Huggins, Charles, and Philip Johnson Clark. 1940. Quantitative Studies of Prostatic Secretion. *The Journal of Experimental Medicine* 72: 747–762.

Huggins, Charles, and Clarence Hodges. 1940. Studies on Prostatic Cancer: I. The Effect of Castration, of Estrogen and of Androgen Injection on Serum Phosphatases in Metastatic Carcinoma of the Prostate. *The Journal of Urology* 167: 948–951.

Huggins, Charles, M.H. Masina, Lillian Eichelberger, and James D. Wharton. 1939. Quantitative Studies of Prostatic Secretion. *The Journal of Experimental Medicine* 70: 543–556.

Huggins, Charles, R.E. Stevens, and Clarence Hodges. 1941. Studies on Prostatic Cancer: II. The Effects of Castration on Advanced Carcinoma of the Prostate Gland. *Archives of Surgery* 43: 209–223.

Huggins, Charles, and R.E. Stevens. 1940. The Effect of Castration on Benign Hypertrophy of the Prostate in Man. *Journal of Urology* 43: 705–714.

Hunter, John. 1792. *Observations on Certain Parts of the Animal Oeconomy*. London: G. Nichol.

———. 1788. *A Treatise on the Venereal Disease*. London: G. Nichol.

James, Nicholas. 2011. *Cancer: A Very Short Introduction*. Oxford: Oxford University Press.

Jewett, H.J. 1975. The Present Status of Radical Prostatectomy for Stages A and B Prostatic Cancer. *The Urologic Clinics of North America* 2: 105–124.

Johnson, Steven. n.d. House Committee Passes HHS Funding Bill That Ends AHRQ. *Modern Healthcare*. June 25, 2015.

Johnstone, Emm Barnes, and Joanna Baines. 2015. *The Changing Faces of Childhood Cancer: Clinical and Cultural Visions since 1940*. Basingstoke: Palgrave Macmillan.

Jolly, Jacques. 1869. *Essai Sur le Cancer de la Prostate*. Paris: P. Asselin.

Jones, Bleddyn. n.d. The Neutron-Therapy Saga: A Cautionary Tale. MedicalPhysicsWeb, January 17, 2008. Accessed May 29, 2015. http://medicalphysicsweb.org/cws/article/opinion/32466

Jones, Mark Peter. 2011. Networked Success and Failure at Hybritech. *Bulletin of Science, Technology & Society* 31: 448–459.

Marx, Franz Josef, and Axel Karenberg. 2009. History of the Term Prostate. *The Prostate* 69: 208–213.

Justman, Stewart. 2010. How Did the PSA System Arise? *Journal of the Royal Society of Medicine* 103: 309–312.

———. 2011. What's Wrong With Chemoprevention of Prostate Cancer? *American Journal of Bioethics* 11: 21–25.

———. 2015. *The Nocebo Effect: Overdiagnosis and Its Costs*. New York: Palgrave Macmillan.

Kacperek, A. March 2009. Proton Therapy of Eye Tumours in the UK: A Review of Treatment at Clatterbridge. *Applied Radiation and Isotopes* 67: 378–386.

Robert A. Kagan, and Robert J. Schulz. 2010. Proton-Beam Therapy for Prostate Cancer. *Cancer* 16: 405–409.

Kammer, Jerry. n.d. Lobbying Ties Give Campus Funding Edge. *The San Diego Union-Tribune*, December 23, 2005. Accessed March 19, 2015. http://www. utsandiego.com/uniontrib/20051223/news_1n23lewisid1.html

———. n.d. A Steady Flow of Financial Influence. *The San Diego Union-Tribune*, December 23, 2005. Accessed September 7, 2015. http://www.pulitzer.org/archives/7051

Keating, Peter, and Alberto Cambrosio. 2012. *Cancer on Trial: Oncology as a New Style of Practice*. Chicago: The University of Chicago Press.

Keating, Peter. 2002. From Screening to Clinical Research: The Cure of Leukemia and the Early Development of the Cooperative Oncology Groups, 1955–1966. *Bulletin of the History of Medicine* 76: 299–334.

Kenyon, Herbert. 1950. *The Prostate Gland*. New York: Random House.

King, Samantha. 2006. *Pink Ribbons, Inc.: Breast Cancer and the Politics of Philanthropy*. Minneapolis: University of Minnesota Press.

Kolata, Gina. n.d. How Demand Surged for Prostate Test. *New York Times*, September 29, 1993. Accessed September 12, 2015. http://www.nytimes.com/1993/09/29/health/how-demand-surged-for-prostate-test.html

Kramer, Barnett, and Jennifer Miller Croswell. 2009. Cancer Screening: The Clash of Science and Intuition. *Annual Review of Medicine* 60: 125–137.

Kuriyama, M., M.C. Wang, L.D. Papsidero, C.S. Killian, T. Shimano, L. Valenzuela, T. Nishiura, G.P. Murphy, and T.M. Chu. 1980. Quantitation of Prostate-Specific Antigen in Serum by a Sensitive Enzyme Immunoassay. *Cancer Research* 40: 4658–4662.

Kutcher, Gerald. 2009. *Contested Medicine Cancer Research and the Military*. Chicago: University of Chicago Press.

Lau, Gloria. n.d. U.S. Oncology Inc. Houston, Texas Medical Firm Moves Ahead In Tough Market. *Investor's Business Daily*, March 26, 2001. Accessed September 27, 2015. http://news.investors.com/business-the-new-america/032601-343453-us-oncology-inc-houston-texas-medical-firm-moves-ahead-in-tough-market.htm

Lawrence, Christopher. 1981. Medicine The Genesis of Cancer: A Study in the History of Ideas. *The British Journal for the History of Science* 14: 210–211.

Lawrence, William. 1817. Cases of Fungus Hæmatodes, with Observations, by George Langstaff, Esq. and an Appendix, Containing Two Cases of Analogous Affections. *Medico-Chirurgical Transactions* 8: 272–314.

Hunter, John. 1839. *Lectures on the Principles of Surgery*. London: Barrington and Haswell.

Lederer, Susan E. 1997. *Subjected to Science: Human Experimentation in America before the Second World War*. Baltimore: Johns Hopkins University Press.

Lee, Jaimey. n.d. As a Proton Therapy Center Closes, Some See It as a Sign. *Modern Healthcare*, September 18, 2014. Accessed September 25, 2015. http://www.modernhealthcare.com/article/20140918/NEWS/309189939

Lee, R. Alton. 2002. *The Bizarre Careers of John R. Brinkley.* Lexington: University Press of Kentucky.

Lenzer, Jeanne. 2013. Professor Who Criticized Prostate Screening Seminar Did Not Suffer Retaliation, Says University. *British Medical Journal* 346: 327.

———. 2012. Professor Was Harassed by His University After Criticising Routine Prostate Cancer Screening, Inquiry Finds. *British Medical Journal* 344: 4150.

Leonhardt, David. n.d. Health Reform's Acid Test: Prostate Cancer'. *The New York Times,* July 7, 2009. Accessed 25 September 2015. http://www.nytimes.com/2009/07/08/business/economy/08leonhardt.html

Lerner, Barron. 2002. Breast Cancer Activism: Past Lessons, Future Directions. *Nature Reviews Cancer* 225–30.

Lewis, Bransford. 1908. The Dawn and Development of Urology. *Transactions of the American Urological Association* 1: 9–20.

Lewis, B. 2008. On Equipoise and Emerging Technologies. *Journal of Clinical Oncology* 15: 2590.

Lexchin, Joel. 2012. Those Who Have the Gold Make the Evidence: How the Pharmaceutical Industry Biases the Outcomes of Clinical Trials of Medications. *Science and Engineering Ethics* 18: 247–261.

London Health Science Center. n.d. Celebrating the 60th Anniversary of the World's First Cancer Treatment with Cobalt-60 Radiation. Accessed September 26, 2015. http://www.lhsc.on.ca/About_Us/LHSC/Publications/Features/Cobalt-60.htm

Löwy, Ilana. 2011. *A Woman's Disease: The History of Cervical Cancer.* Oxford: Oxford University Press.

———. 1996. *Between Bench and Bedside: Science, Healing, and Interleukin-2 in a Cancer Ward.* Cambridge: Harvard University Press.

Machtens, S., D. Schultheiss, M. Kuczyk, M.C. Truss, and U. Jonas. 2000. The History of Endocrine Therapy of Benign and Malignant Diseases of the Prostate. *World Journal of Urology* 18: 222–226.

Mackie, T.R. 2006. History of Tomotherapy. *Physics in Medicine and Biology* 51: R427–R453.

Marks, Harry. 2000. *The Progress of Experiment: Science and Therapeutic Reform in the United States, 1900–1990.* Cambridge: Cambridge University Press.

Maugh, Thomas. 2012. *Seeds of Destruction: The Science Report on Cancer Research.* New York: Springer.

Mazzitelli, Giovanni. 2011. Application of Particle Accelerators in Research. *Radiation Protection Dosimetry* 146: 372–376.

McCallum, Jack. 2013. *The Prostate Monologues: What Every Man Can Learn from My Humbling, Confusing, and Sometimes Comical Battle with Prostate Cancer.* Emmaus: Rodale.

Medical Research Council. 1948. Streptomycin Treatment of Pulmonary Tuberculosis. *British Medical Journal* 2: 769–782.

Mellinger, George, Donald Gleason, and John Bailar. 1967. The Histology and Prognosis of Prostatic Cancer. *The Journal of Urology* 97: 331–337.

Mercier, Louis. 1856. *Recherches Sur Le Traitement Des Maladies Des Organes Urinaires*. Paris: Labe.

Mettlin, C., and Gerald Murphy. 1981. Cancer among Black Populations. Proceedings of the International Conference on Cancer among Black Populations, Buffalo, New York, May 5–6, 1980. *Progress in Clinical and Biological Research* 53: 1–271.

Miele, Rachelle, and Juanne Clarke. 2014. "We Remain Very Much the Second Sex": The Constructions of Prostate Cancer in Popular News Magazines, 2000–2010. *American Journal of Men's Health* 8: 15–25.

Millin, Terence. 1945. Retropubic Prostatectomy: A New Extravesical Technique. *Lancet* 249: 693–696.

Mitchell, J.P. 1970. Transurethral Resection. *British Medical Journal* 3: 241–246.

Moog, Ferdinand Peter, Axel Karenberg, and Friedrich Moll. 2005. The Catheter and its use from Hippocrates to Galen. *The Journal of Urology* 174: 1196–1198.

Moore, Wendy. 2007. *The Knife Man: Blood, Body Snatching, and the Birth of Modern Surgery*. New York: Broadway Books.

Morgagni, Giambattista, and William Cooke. 1824. *The Seats and Causes of Diseases: Investigated by Anatomy*, vol 2. London: Wells and Lilly.

Morris, Zoë Slote, Steven Wooding, and Jonathan Grant. 2011. The Answer Is 17 Years, What Is the Question: Understanding Time Lags in Translational Research. *Journal of the Royal Society of Medicine* 104: 510–520.

Movember. n.d. Movember United States. Accessed October 1, 2015. https://us.movember.com/?home

Mumford, Kevin J. 1992. "Lost Manhood" Found: Male Sexual Impotence and Victorian Culture in the United States. *Journal of the History of Sexuality* 3: 33–57.

Murphy, Gerald. 1999. Review of Phase II Hormone Refractory Prostate Cancer Trials. *Urology* 54: 19–21.

———. 1974. Prostate Cancer. *CA: A Cancer Journal for Clinicians* 24: 282–288.

Murphy, Gerald, and W.F. Whitmore. 1979. A Report of the Workshops on the Current Status of the Histologic Grading of Prostate Cancer. *Cancer* 44: 1490–1494.

Murphy, Gerald, J.T. Leonard, and Ernest Desnos. 1972. *The History of Urology*. Springfield: Thomas.

Nass, Sharyl, Harold Moses, and John Mendelson. 2010. *A National Cancer Clinical Trials System for the 21st Century: Reinvigorating the NCI Cooperative Group Program*. Washington, D.C.: National Academies Press.

National Association for Proton Therapy. n.d. Community Forum. Accessed September 25, 2015. http://www.proton-therapy.org/forum.htm

Nesbit, R.M., and W.C. Baum. 1950. Endocrine Control of Prostatic Carcinoma; Clinical and Statistical Survey of 1,818 Cases. *Journal of the American Medical Association* 143: 1317–1320.

Nutton, Vivian. 2012. *Ancient Medicine*. New York: Routledge.

Olsen, D.R., O.S. Bruland, G. Frykholm, and I.N. Norderhaug. 2007. Proton Therapy. A Systematic Review of Clinical Effectiveness. *Radiotherapy and Oncology* 83: 123–132.

Olson, James. 2009. *Making Cancer History: Disease and Discovery at the University of Texas M.D. Anderson Cancer Center*. Baltimore: Johns Hopkins University Press.

O'Shea, Christopher. 2012. "A Plea for the Prostate": Doctors, Prostate Dysfunction, and Male Sexuality in Late 19th- and Early 20th-Century Canada. *Canadian Bulletin of Medical History* 29: 7–27.

Otis, Laura. 2007. *Müller's Lab*. Oxford: Oxford University Press.

Oudshoorn, Nelly. 1994. *Beyond the Natural Body: An Archaeology of Sex Hormones*. London: Routledge.

Overall, George Whitfield. 1906. *A Non-Surgical Treatise on Diseases of the Prostate Gland and Adnexa*. Chicago: Rowe.

Parascandola, J. 1981. The Theoretical Basis of Paul Ehrlich's Chemotherapy. *Journal of the History of Medicine and Allied Sciences* 36: 19.

Parker-Pope, Tara. n.d. Cancer Funding: Does It Add Up?. *New York Times*, March 6, 2008. Accessed October 1, 2015. http://well.blogs.nytimes.com/2008/03/06/cancer-funding-does-it-add-up/

Payne, Lynda E. Stephenson. 2007. *With Words and Knives: Learning Medical Dispassion in Early Modern England*. Aldershot: Ashgate.

Perry, Seymour. 1982. The Brief Life of the National Center for Health Care Technology. *New England Journal of Medicine* 307: 1095–1100.

Phillips, John L., and Akhouri A. Sinha. 2009. Patterns, Art, and Context: Donald Floyd Gleason and the Development of the Gleason Grading System. *Urology* 74: 497–503.

Pickstone, John. 1993. Ways of Knowing: Towards a Historical Sociology of Science, Technology and Medicine. *The British Journal for the History of Science* 26: 433–458.

Porter, Roy. 2002. William Hunter: A Surgeon and a Gentleman. In *William Hunter and the Eigteenth Century Medical World*, ed. William Bynum and Roy Porter, 7–34. Cambridge: Cambridge University Press.

———. 1999. *The Greatest Benefit to Mankind: A Medical History of Humanity*. New York: W.W. Norton.

Potosky, A., B.A. Miller, P.C. Albertsen, and B.S. Kramer. 1995. The Role of Increasing Detection in the Rising Incidence of Prostate Cancer. *Journal of the American Medical Association* 273: 548–552.

Potosky, A.L., L. Kessler, G. Gridley, C.C. Brown, and J.W. Horm. 1990. Rise in Prostatic Cancer Incidence Associated with Increased Use of Transurethral Resection. *Journal of the National Cancer Institute* 82: 1624–1628.

Potts, Jeannette, and Esteban Walker. 2010. Isn't It Time to Abandon Prostate Specific Antigen (PSA) for Prostate Cancer Screening? *Journal of Men's Health* 7: 320.

ProCure. n.d. Learn About ProCure State of the Art Cancer Treatment Centers | Locations | ProCure Proton Therapy.' Accessed September 27, 2015. https://www.procure.com/Locations

PTCOG. n.d. *Particles*. Newsletters. Accessed September 26, 2015. http://www.ptcog.ch/index.php/particles-newsletters

Rasmussen, Nicolas. 2004. The Moral Economy of the Drug Company-Medical Scientist Collaboration in Interwar America. *Social Studies of Science* 34: 161–185.

Rather, L.J. 1978. *The Genesis of Cancer: A Study in the History of Ideas*. Baltimore: Johns Hopkins University Press.

Reiser, Stanley J. 1988. *Medicine and the Reign of Technology*. Cambridge: Cambridge University Press.

——— 2014. *Technological Medicine: The Changing World of Doctors and Patients*. Cambridge: Cambridge University Press.

Rich, Arnold Rice. 2007. On the Frequency of Occurrence of Occult Carcinoma of the Prostrate. *International Journal of Epidemiology* 36: 274–277.

Riches, Eric. 1968. The History of Lithotomy and Lithotrity. *Annals of the Royal College of Surgeons of England* 43: 185–199.

Ricketts, B. Merrill. 1904. *Surgery of the Prostate, Pancreas, Diaphragm, Spleen, Thyroid, and Hydrocephalus: A Historical and Bibliographical Review*. Cincinnati.

Ries, L., M. Eisner, C. Kosary, B. Hankey, B. Miller, L. Clegg, and B. Edwards, eds. 1994. *SEER Cancer Statistics Review, 1973–1991*. Bethesda: National Cancer Institute.

Rosenberg, Charles. 2009. Managed Fear. *Lancet* 373: 802–803.

———. 2007. *Our Present Complaint: American Medicine, Then and Now*. Baltimore: Johns Hopkins University Press.

Rosenman, Ellen Bayuk. 2003. Body Doubles: The Spermatorrhea Panic. *Journal of the History of Sexuality* 12: 365–399.

Rosevear, H.M., A.J. Lightfoot, M.A. O'Donnell, C.E. Platz, S.A. Loening, and C.E. Hawtrey. 2011. Rubin H. Flocks and Colloidal Gold Treatments for Prostate Cancer. *The Scientific World Journal* 11: 1560–1567.

Rothman, David. 1991. *Strangers at the Bedside: A History of How Law and Bioethics Transformed Medical Decision Making*. New York: Basic Books.

Roth, Russell, Elmer Hess, and Anthony Kaminsky. 1954. Of Fires and Frying Pans. *Journal of the American Medical Association* 155: 302.

Ruck, Sean. n.d. Q&A with Leonard Arzt: Founder and Past Executive Director of the National Association for Proton Therapy. *Dotmed.com*. Accessed September 25, 2015. http://www.dotmed.com/news/story/25151

Rutten, Thomas. 2011. Early Modern Medicine.' In *The Oxford Handbook of the History of Medicine*, edited by Mark Jackson 60–81. Oxford: Oxford University Press.

Sackett, David, W.M. Rosenberg, J.A. Gray, R.B. Haynes, and W.S. Richardson. 1996. Evidence Based Medicine: What It Is and What It Isn't. *British Medical Journal* 312: 71–72.

Sakr, W., D.J. Grignon, G.P. Haas, L.K. Heilbrun, J.E. Pontes, and J.D. Crissman. 1996. Age and Racial Distribution of Prostatic Intraepithelial Neoplasia. *European Urology* 30: 138.

Sanders, Mark. 2005. William Cowper and His Decorated Copperplate Initials. *Anatomical Record* 282: 5–12.

Schröder, F.H., J. Hugosson, M.J. Roobol, T.L. Tammela, S. Ciatto, V. Nelen, M. Kwiatkowski, et al. 2009. Screening and Prostate-Cancer Mortality in a Randomized European Study. *New England Journal of Medicine* 360: 1320–1328.

Schröder, Fritz, J. Hugosson, M.J. Roobol, T.L. Tammela, S. Ciatto, V. Nelen, M. Kwiatkowski, et al. 2012. Prostate-Cancer Mortality at 11 Years of Follow-Up. *The New England Journal of Medicine* 366: 981–990.

Schultheiss, D., J. Denil, and U. Jonas. 1997. Rejuvenation in the Early 20th Century. *Andrologia* 29: 351–355.

Seger, Gordon. 1948. A Cooperative Study of Streptomycin in Tuberculosis. *Chest* 14: 686–693.

Sengoopta, Chandak. 2006. *The Most Secret Quintessence of Life: Sex, Glands, and Hormones, 1850–1950*. Chicago: University of Chicago Press.

———. 2003. "Dr Steinach Coming to Make Old Young!": Sex Glands, Vasectomy and the Quest for Rejuvenation in the Roaring Twenties. *Endeavour* 27: 122–126.

Sheets, Nathan, G.H. Goldin, A.M. Meyer, Y. Wu, Y. Chang, T. Stürmer, J.A. Holmes, et al. 2012. Intensity-Modulated Radiation Therapy, Proton Therapy, or Conformal Radiation Therapy and Morbidity and Disease Control in Localized Prostate Cancer. *Journal of the American Medical Association* 307: 1611–1620.

Shelley, H.S. 1969. The Enlarged Prostate. A Brief History of Its Treatment. *Journal of the History of Medicine and Allied Sciences* 24: 452–473.

Showalter, Elaine. 1997. *Hystories: Hysterical Epidemics and Modern Culture*. London: Picador.

Slater, J., J.O. Archambeau, D.W. Miller, M.I. Notarus, W. Preston, J.D. Slater, et al. 1992. The Proton Treatment Center at Loma Linda University Medical

Center: Rationale for and Description of Its Development. *International Journal of Radiation Oncology, Biology, Physics* 22: 383–389.

Smith, Alfred. 2009. Vision 20/20: Proton Therapy. *Medical Physics* 36: 556–568.

Soanes, W., R. Ablin, and M. Gonder. 1970. Remission of Metastatic Lesions Following Cryosurgery in Prostatic Cancer: Immunologic Considerations. *The Journal of Urology* 104: 154–159.

Srigley, J., M. Amin, L. Boccon-Gibod, L. Egevad, J.I. Epstein, P.A. Humphrey, G. Mikuz, et al. 2005. Prognostic and Predictive Factors in Prostate Cancer: Historical Perspectives and Recent International Consensus Initiatives. *Scandinavian Journal of Urology and Nephrology* 216: 8–19.

Stamey, T., M. Caldwell, J.E. McNeal, R. Nolley, M. Hemenez, and J. Downs. 2004. The Prostate Specific Antigen Era in the United States is Over for Prostate Cancer: What Happened in the Last 20 Years? *The Journal of Urology* 172: 1297–1301.

Stamey, T., J.N. Kabalin, J.E. McNeal, I.M. Johnstone, F. Freiha, E.A. Redwine, N. Yang, et al. 1989. Prostate Specific Antigen in the Diagnosis and Treatment of Adenocarcinoma of the Prostate. IV. Anti-Androgen Treated Patients. *The Journal of Urology* 141: 1088–1090.

Stamey, T., N. Yang, A.R. Hay, J.E. McNeal, F.S. Freiha, and E. Redwine. 1987. Prostate-Specific Antigen as a Serum Marker for Adenocarcinoma of the Prostate. *New England Journal of Medicine* 317: 909–916.

Stanley, L. 1931. Testicular Substance Implantation: Comments on some Six Thousand Implantations. *California and Western Medicine* 35: 411–415.

Stark, Laura. 2012. *Behind Closed Doors: IRBs and the Making of Ethical Research.* Chicago: The University of Chicago Press.

Starling, Ernest. 1905. On the Chemical Correlation of the Functions of the Body. *Lancet* 166: 501–503.

Starr, Paul. 1982. *The Social Transformation of American Medicine.* New York: Basic Books.

Stephens, Elizabeth. 2008. Pathologizing Leaky Male Bodies: Spermatorrhea in Nineteenth-Century British Medicine and Popular Anatomical Museums. *Journal of the History of Sexuality* 17: 421–438.

Stone, Chester. 1935. *The Dangerous Age in Men: A Treatise on the Prostate Gland.* New York: The Macmillan Company.

Stone, Robert, and John Larkin. 1942. The Treatment of Cancer with Fast Neutrons. *Radiology* 39: 608–620.

Streptomycin Conference, United States, and Veterans Administration. 1950. *Transactions of the Streptomycin Conference.* Veterans Administration: Washington.

Suit, H., H. Kooy, A. Trofimov, J. Farr, J. Munzenrider, T. DeLaney, J. Loeffler, et al. 2008. Should Positive Phase III Clinical Trial Data Be Required before

Proton Beam Therapy Is More Widely Adopted? No. *Radiotherapy and Oncology* 86: 148–153.

Szabo, Lisa. n.d. Urology Group Stops Recommending Routine PSA Test'. *USA TODAY*, May 3, 2013. Accessed August 15, 2015. http://www.usatoday.com/story/news/nation/2013/05/03/urologists-psa-screening/2130971/

Terasawa, T., T. Dvorak, S. Ip, G. Raman, J. Lau, and T.A. Trikalinos. 2009. Systematic Review: Charged-Particle Radiation Therapy for Cancer. *Annals of Internal Medicine* 151: 556–565.

Thomas, Lewis. 1988. On the Science and Technology of Medicine. *Daedalus* 117: 299–316.

Thompson, Henry. 1873. *The Diseases of the Prostate, Their Pathology and Treatment.* London: J. and A. Churchill.

Thompson, I., E. Canby-Hagino, and M.S. Lucia. 2005. Stage Migration and Grade Inflation in Prostate Cancer: Will Rogers Meets Garrison Keillor. *Journal of the National Cancer Institute* 97: 1236–1237.

Thompson, I., P.J. Goodman, C.M. Tangen, M.S. Lucia, G.J. Miller, L.G. Ford, M.M. Lieber, et al. 2003. The Influence of Finasteride on the Development of Prostate Cancer. *New England Journal of Medicine* 349: 215–224.

Timmermann, Carsten. 2014. *A History of Lung Cancer: The Recalcitrant Disease.* Basingstoke: Palgrave Macmillan.

Trikalinos, T., T. Terasawa, S. Ip, G. Raman, and J. Lau. 2009. *Particle Beam Radiation Therapies for Cancer. Technical Brief No. 1.* Rockville, MD: Agency for Healthcare Research and Quality.

Ueyama, Takahiro, and Christopher Lecuyer. 2006. Building Science-Based Medicine at Stanford: Henry Kaplan and the Medical Linear Accelerator, 1948–1975'. In *Devices and Designs: Medical Technologies in Historical Perspective*, edited by Carsten Timmermann and Julie Anderson, 137–155. Basingstoke: Palgrave Macmillan.

US Preventive Services Task Force. n.d. Final Update Summary: Breast Cancer: Screening. November 2009. Accessed September 3, 2015. http://www.uspreventiveservicestaskforce.org/Page/Document/UpdateSummaryFinal/breast-cancer-screening?ds=1&s=mammography

US Preventive Services Task Force. n.d. Final Update Summary: Prostate Cancer: Screening. May 2012. Accessed September 3, 2015. http://www.uspreventiveservicestaskforce.org/Page/Document/UpdateSummaryFinal/prostate-cancer-screening

Veldeman, L., I. Madani, F. Hulstaert, G. De Meerleer, M. Mareel, and W. De Neve. 2008. Evidence Behind Use of Intensity-Modulated Radiotherapy: A Systematic Review of Comparative Clinical Studies. *The Lancet. Oncology* 9: 367–375.

Von Staden, Hienrich. 1995. Anatomy as Rhetoric: Galen on Dissection and Persuasion. *Journal of the History of Medicine and Allied Sciences* 50: 47–66.

Vos, Rein. 1989. *Drugs Looking for Diseases Innovative Drug Research and the Development of the Beta Blockers and the Calcium Antagonists.* Dordrecht: Springer Netherlands.

Wailoo, Keith. 2011. *How Cancer Crossed the Color Line.* Oxford: Oxford University Press.

Walsh, Patrick C. 2012. How Charles Huggins Made His Nobel Prize Winning Discovery—in His Own Words: An Historic Audio Recording. *The Prostate* 72: 1718.

Wang, M.C., L.A. Valenzuela, G.P. Murphy, and T.M. Chu. 1979. Purification of a Human Prostate Specific Antigen. *Investigative Urology* 17: 159–163.

Warner, John Harley. 1997. From Specificity to Universalism in Medical Therapeutics: Transformation in the 19th-Century United States. In *Sickness and Health in America: Readings in the History of Medicine and Public Health*, eds. Judith Leavitt and Ronald Numbers, 87–101. Madison: University of Wisconsin Press.

Wassersug, Richard, John Oliffe, and Christina Han. 2015. On Manhood and Movember ... or Why the Moustache Works. *Global Health Promotion* 22: 65–70.

Wenner, Melinda. n.d. The War against War Metaphors *The Scientist*, February 16 2007. Accessed September 30, 2015. http://www.the-scientist.com/?articles.view/articleNo/24756/title/The-war-against-war-metaphors/

Wegwarth, Odette, Lisa M. Schwartz, Steven Woloshin, Wolfgang Gaissmaier, and Gerd Gigerenzer. 2012. Do Physicians Understand Cancer Screening Statistics? A National Survey of Primary Care Physicians in the United States. *Annals of Internal Medicine* 156: 340–349.

Weisz, George. 2006. *Divide and Conquer: A Comparative History of Medical Specialization.* Oxford: Oxford University Press.

Welch, H. Gilbert, Lisa M. Schwartz, and Steven Woloshin. 2011. *Overdiagnosed: Making People Sick in the Pursuit of Health.* Boston: Beacon Press.

White, William. 1895. I. The Results of Double Castration in Hypertrophy of the Prostate. *Annals of Surgery* 22: 1–80.

Whitmore, W.F. 1956. Hormone Therapy in Prostatic Cancer. *The American Journal of Medicine* 21: 697–713.

Woloshin, S., and L.M. Schwartz. 1999. The U.S. Postal Service and Cancer Screening–Stamps of Approval? *The New England Journal of Medicine* 340: 884–887.

Woodward, W., E.A. Strom, S.L. Tucker, M.D. McNeese, G.H. Perkins, N.R. Schechter, S.E. Singletary, et al. 2003. Changes in the 2003 American Joint Committee on Cancer Staging for Breast Cancer Dramatically Affect Stage-Specific Survival. *Journal of Clinical Oncology* 17: 3244–3248.

Yallop, Helen. 2015. *Age and Identity in Eighteenth-Century England.* Hoboken: Taylor and Francis.

Yamey, Gavin, and Michael Wilkes. 2002. The PSA Storm. *British Medical Journal* 324: 431.

Yates, David R., Christophe Vaessen, and Morgan Roupret. 2011. From Leonardo to Da Vinci: The History of Robot-Assisted Surgery in Urology. *British Journal of Urology International* 108: 1708–1713.

Yoshioka, Alan. 1998. Use of Randomisation in the Medical Research Council's Clinical Trial of Streptomycin in Pulmonary Tuberculosis in the 1940s. *British Medical Journal* 317: 1220–1223.

Young, H. 1910. Perineal Prostatectomy: Some Conclusions Based on a Study of 400 Cases. *Journal of the American Medical Association* 10: 784–791.

———. 1935. Radical Cure of Carcinoma of the Prostate. *The American Journal of Surgery* 28: 32–46.

———. 1922. *Technique of Radium Treatment of Cancer of the Prostate Gland and Seminal Vesicles /*. [S.l. : http://hdl.handle.net/2027/osu.32435001185644

Zubrod, Gordon. 1984. Origins and Development of Chemotherapy Research at the National Cancer Institute. *Cancer Treatment Reports* 68: 9–19.

INDEX

© The Editor(s) (if applicable) and The Author(s) 2016   229
H. Valier, *A History of Prostate Cancer*,
DOI 10.1057/978-1-137-56595-2